点线间融合中西建筑科技，延续百年岭南现代建筑设计神韵

The Dotted Line Fusion of Chinese and Western Architectural Technology, Continuation of Centuries Lingnan Genre of Architectural Design Charm.

筑梦天下 黄劲
Architectural Dreams of Huang Jin

新岭南建筑师专辑
The Architect of South China (Lingnan) Series

黄 劲　黄思宇 编著　　叶 飚 摄影
By Huang Jin, Huang Siyu　　Photography By Billy Yip

辽宁科学技术出版社

关于黄劲
ABOUT HUANG JIN

1966年6月出生于广州（原籍佛山三水）	Born in Guangzhou in June 1966 (The Ancestral Home is in Sanshui of Foshan)
1984年毕业于广州市执信中学	Graduated from Zhixin High School of Guangzhou in 1984
1988年毕业于华南理工大学建筑系建筑学专业，本科学历	Graduated from School of Architecture of South China University of Technology with Bachelor degree in 1988
教授级高级建筑师	Senior Architect of Professor level
国务院特殊津贴专家	State-Council Allowance Obtained Expert
国家一级注册建筑师	National First-Class Architect
资深规划设计师	Senior Planner
资深室内设计师	Senior Interior Designer
资深园林设计师	Senior Landscape Designer

个人主要作品：

[1] 华南第一学府中山大学：《建筑创作》。天津：天津大学出版社
[2] 南海市枫丹白露酒店。《建筑学报》
[3] 广州购书中心。《建筑学报》
[4] 广州大学城中山大学图书馆。《建筑学报》
[5] 广州大学城组团——规划与建筑设计。《建筑学报》
[6] 越地域，越国际。第一届全国建筑创作设计高峰论坛讲稿
[7] 黄劲，中国青年建筑师·当代中国新作品：《建筑创作》

Key Works:

[1] The First Institution of South China: Sun Yat-sen University: Editor of ArchiCreation Magazine. Tianjin: Tianjin University Press
[2] Fontainebleau Hotel of Nanhai. Journal of Architecture
[3] Guangzhou Book Center. Journal of Architecture
[4] Library of the Sun Yat-sen University in Guangzhou Higher Education Mega Center. Journal of Architecture
[5] Planning and Architecture Design of Guangzhou Higher Education Mega Center. Journal of Architecture
[6] Regional is International. Speech on the Summit Forum of the First National Architectural Creation Design
[7] Huangjing. Chinese Young Architec. Contemporary Chinese New Design: ArchiCreation Magazine

序号 Serial Number	项目名称 Project Name	曾获奖项 Awards
01	广州大学城组团一的规划设计（中山大学和广东外语外贸大学） The First Group Participation-Guangzhou Higher Education Mega Center(Sun Yat-sen University & Guangdong University of Foreign Studies)	① 国际竞赛第一名 ② 广东省优秀建筑设计一等奖 ③ 广东省优秀规划设计二等奖 ④ 住房和城乡建设部建筑学会创作奖·佳作奖 ⑤ 住房和城乡建设部优秀建筑设计三等奖 ① First Prize, International Competition ② First Prize, Excellent Architectural Design of Guangdong Province ③ Second Prize, Excellent Planning & Design of Guangdong Province ④ Honorable Mention Award, Architectural Society of China Architecture Creative Awards By Ministry of Construction ⑤ Third Prize, Excellent Design of Ministry of Construction
02	津滨·腾越大厦 Jinbin·Tengyue Mansion	① 住房和城乡建设部优秀建筑设计二等奖 ② 广东省优秀建筑设计二等奖 ① Second Prize, Excellent Design of Ministry of Construction ② Second Prize, Excellent Architectural Design of Guangdong Province
03	勤建大厦 Qinjian Building	广东省优秀建筑设计三等奖 Third Prize, Excellent Architectural Design of Guangdong Province
04	广州购书中心与维多利广场 Guangzhou Book Center & Victory Plaza	① 广东省优秀建筑设计一等奖 ② 住房和城乡建设部优秀建筑设计二等奖 ③ 国家优秀设计铜奖 ① First Prize, Excellent Architectural Design of Guangdong Province ② Second Prize, Excellent Design of Ministry of Construction ③ Bronze Award, State Excellent Design
05	新华大厦 Xinhua Building	广东省优秀建筑设计二等奖 Second Prize, Excellent Architectural Design of Guangdong Province
06	佛山市南海区国土局地籍资料库 Cadastral Database of Land and Resources Bureau of Nanhai District of Foshan	广东省优秀建筑设计三等奖 Third Prize, Excellent Design of Guangdong Province
07	松岗广场 Songgang Square	① 住房和城乡建设部优秀建筑设计三等奖 ② 广东省优秀建筑设计二等奖 ① Third Prize, Excellent Design of Ministry of Construction ② Second Prize, Excellent Design of Guangdong Province
08	佛山市南海区国家税务局综合业务用房 Integrated Service Office of State Administration of Taxation of Nanhai District of Foshan	广东省优秀建筑设计二等奖 Second Prize, Excellent Architectural Design of Guangdong Province
09	老挝亚欧首脑峰会接待大酒店 Landmark Meckong Riverside Hotel of Asia-Europe Meeting Summit Hotel, Laos	国际竞赛中标 Bid Winner of International Design Competition
10	老挝万象广晟大酒店 Guangsheng Hotel, Vientiane, Laos	国际竞赛中标 Bid Winner of International Design Competition
11	广州云来斯堡酒店、办公综合楼 Vanburgh Hotel, Office Building	① 设计竞赛中标 ② 2014年度全国工程建设项目优秀设计成果二等奖 ① Bid Winner of Design ② Second Prize, 2014 Annual National Construction Project Design Excellence Awarded
12	佛山市南国桃园枫丹白鹭酒店 Fontainebleau Hotel, Nanguo Peach Garden, Foshan	广东省优秀建筑设计二等奖 Second Prize, Excellent Architectural Design of Guangdong Province
13	中山大学东校区图书馆 Library of the East Campus, Sun Yat-sen University	① 广东省优秀建筑设计一等奖 ② 住房和城乡建设部优秀建筑设计三等奖 ③ 住房和城乡建设部建筑学会创作奖·佳作奖 ① First Prize, Excellent Architectural Design of Guangdong Province ② Third Prize, Excellent Design of Ministry of Construction ③ Honorable Mention Award, Architectural Society of China Architecture Creative Awards, by Ministry of Construction

序号 Serial Number	项目名称 Project Name	曾获奖项 Awards
14	中山大学东校区公共教学楼 Public Classroom Building of East Campus, Sun Yat-sen University	① 广东省优秀建筑设计一等奖 ② 住房和城乡建设部优秀建筑设计三等奖 ③ 住房和城乡建设部建筑学会创作奖·佳作奖 ① First Prize, Excellent Architectural Design of Guangdong Province ② Third Prize, Excellent Design of Ministry of Construction ③ Honorable Mention Award, Architectural Society of China Architecture Creative Awards, by Ministry of Construction
15	中山大学东校区法学院 School of Low of East Campus, Sun Yat-sen University	① 广东省优秀建筑设计一等奖 ② 住房和城乡建设部优秀建筑设计三等奖 ③ 住房和城乡建设部建筑学会创作奖·佳作奖 ① First Prize, Excellent Architectural Design of Guangdong Province ② Third Prize, Excellent Design of Ministry of Construction ③ Honorable Mention Award, Architectural Society of China Architecture Creative Awards, by Ministry of Construction
16	中山大学东校区行政学院 School of Administration of East Campus, Sun Yat-sen University	① 广东省优秀建筑设计一等奖 ② 住房和城乡建设部优秀建筑设计三等奖 ③ 住房和城乡建设部建筑学会创作奖·佳作奖 ① First Prize, Excellent Architectural Design of Guangdong Province ② Third Prize, Excellent Design of Ministry of Construction ③ Honorable Mention Award, Architectural Society of China Architecture Creative Awards, by Ministry of Construction
17	中山大学东校区传播与设计学院 School of Communications and Design of East Campus, Sun Yat-sen University Communications and Design of East Campus, Sun Yat-sen University	① 广东省优秀建筑设计一等奖 ② 住房和城乡建设部优秀建筑设计三等奖 ③ 住房和城乡建设部建筑学会创作奖·佳作奖 ① First Prize, Excellent Architectural Design of Guangdong Province ② Third Prize, Excellent Design of Ministry of Construction ③ Honorable Mention Award, Architectural Society of China Architecture Creative Awards, by Ministry of Construction
18	中山大学东校区基础实验楼 Fundamental Laboratory Building of East Campus, Sun Yat-sen University	① 广东省优秀建筑设计一等奖 ② 住房和城乡建设部优秀建筑设计三等奖 ③ 住房和城乡建设部建筑学会创作奖·佳作奖 ① First Prize, Excellent Architectural Design of Guangdong Province ② Third Prize, Excellent Design of Ministry of Construction ③ Honorable Mention Award, Architectural Society of China Architecture Creative Awards, by Ministry of Construction
19	中山大学东校区工学院 School of Polytechnics of East Campus, Sun Yat-sen University	① 广东省优秀建筑设计一等奖 ② 住房和城乡建设部优秀建筑设计三等奖 ③ 住房和城乡建设部建筑学会创作奖·佳作奖 ① First Prize, Excellent Architectural Design of Guangdong Province ② Third Prize, Excellent Design of Ministry of Construction ③ Honorable Mention Award, Architectural Society of China Architecture Creative Awards, by Ministry of Construction
20	中山大学东校区微纳尺度材料及生命科学实验大楼 Nanoscale Materials & Life Sciences Laboratory Building of East Campus, Sun Yat-sen University	① 广东省优秀建筑设计二等奖 ② 住房和城乡建设部优秀建筑设计三等奖 ① Second Prize, Excellent Architectural Design of Guangdong Province ② Third Prize, Excellent Design of Ministry of Construction
21	中山大学东校区行政会议中心 Administration & Conference Center of East Campus, Sun Yat-sen University	① 广东省优秀建筑设计一等奖 ② 住房和城乡建设部优秀建筑设计三等奖 ① First Prize, Excellent Architectural Design of Guangdong Province ② Third Prize, Excellent Design of Ministry of Construction
22	中山大学东校区工科实验楼与药学院楼 Engineering Laboratory Building & Pharmaceutical Building of East Campus, Sun Yat-sen University	广东省优秀建筑设计二等奖 Second Prize, Excellent Architectural Design of Guangdong Province
23	中山市中国银行大厦 Bank of China Tower, Zhongshan	广东省优秀建筑设计二等奖 Second Prize, Excellent Architectural Design of Guangdong Province
24	广东财经大学华商学院校区 Huashang College, Guangdong University of Finance & Economics	广东省优秀建筑设计三等奖 Third Prize, Excellent Architectural Design of Guangdong Province

作者自序
FOREWORDS

我的建筑观
——越是地域的，就越是世界的
My Architecture Philosophy—Regional is International

黄劲在工作中
Mr. Huang Jin to Work

　　我的少年时代在广州西关大屋中度过，所住西关大屋呈多进院落式结构，其天井的设计具有良好的通风采光，是广州典型的传统居住建筑。而我就读的执信中学和华南理工大学建筑学系，校园建筑均为建国初期的中式建筑佳作。一直在岭南建筑优美环境中熏陶的我，最后报读了建筑学专业。今天回望，这些岭南建筑仍因其独特的地域文化特点让人称道。

　　毕业后这三十年，国内建筑界在不断地探索发展方向。二十年前设计行业就北京建筑风格引发了一场大型学术讨论，意在如何将传统民族风格与现代功能空间结合问题，接着是后现代风格、解构主义等众多思潮的冲击，随后各大城市似乎急于赶超世界水平，引入了众多国际建筑师。国内建筑一直处于借鉴、模仿，甚至拿来的状况。不知不觉中，中国建筑已成外国建筑师的试验场，外国建筑师随手一笔，都让中国建筑师当墨宝看待，不惜造价去实施。

　　反思过去，放眼未来，觉得有必要明确一下自己日后设计的发展方向。根据多年的实践和经验的积累，我提倡建筑设计应走自己的路，走地域设计的路，用现代设计的技术和手法，创造出有鲜明地域特点的新建筑，从而在国际建筑界拥有自己的地位。

　　地域设计包含三重含义：其一是尊重地域自然环境，在建筑采光、通风等方面结合环保、绿色建筑设计理念去创新；其二是尊重地域的人文因素，以人为本，强调在各方面人性化设计，尊重当地居民的生活和民俗习惯；其三是地域的历史文化，建筑已定义为一种文化艺术，我们的作品不能停留在好看好用这些基本层次，我们建筑师应有所追求，力求将作品的内涵上升至文化艺术层面，并且能体现当地文化的特点。

　　地域设计的好处在于有助于我们脱离流行时装的建筑模式，创造出适应当地环境、其有独特文化特点的建筑成果。在全球化浪潮势不可挡的当天，唯有自强自尊，不跟风模仿，具有自己地域特点的设计才能在世界建筑行业拥有自己的一席之位。

　　近年，在地域设计上我做了不少探索，在广州大学城中山大学和广东外语外贸大学设计中，将岭南园林造园手法运用到校园规划之中，在珠江新城津滨腾大厦和勤建大厦中运用了立体绿化、空中花园、高层自然通风等设计，不仅适应了气候环境，而且有很好的节能效果和独特的建筑形象。在老挝亚欧峰会接待酒店的设计中，将老挝文化和法式文化用现代的设计手法融合在一起，得到老挝总理高度赞扬。在设计上重视地域、气候和文化，才能创造出有特色的作品，这种设计方法无国界限制。

　　地域设计它不是一种流派，也不是一种风格，它是一种思想和工作方法，并不受设计手法的限制。每位建筑师都应有表达其创作哲理的个性手法，应运用当地材料和现代手法去创作。同时，建筑师也应有其处理空间和形式的特点。我个人比较偏爱简洁的几何形式，早期设计的购书中心采用三角形，后来的津滨越大厦和新秀大厦等运用了方圆组合，设计上注意去除一些非几何元素，尽力表现出主体鲜明的个性，我希望我的每个作品都具有雕塑美感，成为城市的艺术品。

黄劲

I spent my childhood in Guangzhou Xiguan Mansion. The mansion is with multiple courtyard structure, and the design of "Tianjing (small courtyard)" enables wonderful ventilation and daylight, which represents typical and traditional residential architecture of Guangzhou. In Zhixin High School and Architecture Department of South China University of Technology , I studied in, the buildings on campus are excellent works of Chinese architecture since the initial period of People's Republic of China. Even today such Lingnan architecture is still commendable due to its unique regional and cultural features.

During the thirty years after my graduation, continuous exploration on development direction happened in the architecture field of China. There were large-scale academic discussion regarding Beijing architecture style in the design industry twenty years ago, in the purpose of solving the issue of combination of traditional Chinese style and modern functional space. And then came the impact of various trends of thoughts such as post-modernism and deconstruction. It seemed that major cities in China were in a hurry to surpass international standard, hence a number of foreign architects were introduced. Domestic architecture has been addicted to borrowing, imitating and copying, and slowly and slowly Chinese architecture has become the laboratory for foreign architects, whose casual design was highly valued by Chinese architects and realized at expensive cost.

Reflect on the past and looking to the future, it is necessary to ascertain one's own design direction. According to years of accumulated practice and experience, my advice is that we should always follow our own way, and persist in regional design in architecture design, to create new design in distinct regional style with contemporary design skill and practice, thereby establish our own status in international market.

There are three meanings of regional design: firstly is to respect natural environment, combining design concepts like environmental protection and green architecture in lighting and ventilation to bring forward new idea; secondly is to respect human and culture factors, focusing on various human-based design and value the living habit and custom of local people; thirdly is to pay much attention to local history and culture. Architecture is defined as one type of culture and art, and our design can not linger in the basic levels like being nice and useful. Being an architect, we should pursue more and endeavor to enhance the design connotation in cultural and aesthetic aspects, simultaneously manifest the characteristics of local culture.

Regional design is advantageous to help deviating from popular and fashionable architecture mode, and creating architecture of unique cultural trait adaptable to local environment. In the trend of irresistible globalization today, an architect could only find his own position in the international industry with self-reliance and self-esteem spirit, as well as non-imitating region-oriented design.

I have explored a lot on regional design in recent years. In Sun Yat-sen University and Guangdong University of Foreign Studies in Guangzhou Higher Education Mega Center, approach of Lingnan Garden is applied to the campus planning; in Jinbin Tengyue Building and Qinjian Building in Zhujiang New Town, 3D greening, hanging garden, and naturally ventilation on higher levels not only adapts well to the climate and environment, but also enables excellent energy-saving effect and unique image of architecture; in Asia-Europe Summit Hotel of Laos, Laos culture and French style are integrated by modern design which was highly praised by Laotian Premier. Only by paying much attention to region, climate and culture for design, can an architect create distinctive work, and this is beyond national boundaries.

Regional design is nothing like one type of genre or style, instead, it is a method of thinking and work unlimited by any design practice. Every designer should have his individual method to express his creative philosophy and use indigenous material and modern approach to design, meanwhile possess his own character to manipulate space and form. I personally prefer simple geometric form, such as triangle in Guangzhou Book Center, and later the combination of square and circle in Jinbin Tengyue Building and Xinxiu Building, which some non-geometric elements are intentionally eliminated to outstand the sharp character of the main part. I hope each of my projects could showcase its sculptural beauty and become a piece of art of the city.

Huang Jin

业界与客户的评价
EVALUATION FROM PREES AND CUSTOMERS

为黄劲专辑所写的序
Preface to the Album of Huang Jin

与何镜堂先生（中）和郭卫宏先生在一起
Mr. He Jingtang(betwixt) and Mr. Guo Weihong

何镜堂
He Jingtang

中国工程院院士、中国工程设计大师

华南理工大学建筑学院院长兼设计院院长、教授、博士生导师。国家特许一级注册建筑师、总建筑师、高级建筑师，兼任国务院学位委员会专家评议组成员、全国高等学校建筑专业教育评估委员会委员、广东省科学技术协会副主席、中国建筑学会副理事长、中国建筑学会教育建筑学术委员会主任、广东省注册建筑师协会会长、广东省土木建筑学会副理事长、广东省建筑学会环境艺术学术委员会主任、广州市环境艺术学术委员会副主任、广州市建筑科技委员会副主任、广州市文物管理委员会委员、国家大剧院专家组成员。

Academician of China Academy of Engineering, Design Master of Chinese Engineering

Dean, Professor and Ph.D. Supervisor in School of Architecture, and Director of Architectural Design & Research Institute, South China University of Technology. National First-Class Certified Architect, Chief Architect, Senior Architect, Member of Expert Appraisal Group of Degree Committee of the State Department, Member of Evaluation Committee of Architecture Major Education of National Higher Education Institutions, Vice Chairman of Guangdong Provincial Association of Science and Technology, Vice President of the Architectural Society of China, Director of Academic Committee of Educational Architecture of the Architectural Society of China, President of Guangdong Certified Architect Association, Vice Chairman of the Civil Engineering and Architectural Society of Guangdong, Director of Academic Committee of Environmental Art of Architectural Society of Guangdong, Associate Director of Guangzhou Academic Committee of Environmental Art, Associate Director of Architecture Science and Technology Committee of Guangzhou, Member of Guangzhou Municipal Administration Committee of Cultural Heritage, Member of Expert Group of National Grand Theater of China.

2014年是华南理工大学84届学生从业30周年，适逢在此届的学生利峰博士毕业，与我谈论起他们那几届的学生，在建筑创作上成就甚多，已成为中青年建筑师中的佼佼者，为母校增添了不少光彩。其中利峰提及的在广州市设计院工作20多年的黄劲同学，便是一大批华工培养的优秀建筑师中的一员。

84届学生是首届恢复重点中学后考入华南理工大学的大学生，整体素质比较好。黄劲在学习阶段就有比较突出的表现，其快题作业多次贴堂展示，得到老师同学的认可。1988年学生毕业，广州市设计院钟新权副总建筑师来我校挑选了三位优秀学生，黄劲便是其中的一位。在广州市设计院，黄劲一直师从佘畯南院士从事酒店的创作和研究。

黄劲备受业界瞩目的第一个作品是广州购书中心，在郭明卓大师的指导下，大胆地运用三角形切割手法，构思独特，外形新颖，独具个性。但平面布局却又颇为方正实用，得到评审专家和省市领导的称赞，对于一个刚出茅庐的建筑师，能做出这样的作品，确实是难能可贵。

10年后，黄劲与郭明卓大师合作，在广州大学城中山大学和外语外贸大学的国际方案竞赛中标，而该项目适逢由我和戴蓬院士主持评审，由此对黄劲有了进一步的了解。在设计中黄劲提出利用自然的山水环境，将两所大学以水系相连，从而形成贯通两校的生态文化轴的概念，深得评委会和校方赞赏。在设计中可看出，黄劲十分注重作品的地域性和文化性，并力求融入岭南建筑的特点。

近年黄劲的新作勤建大厦和津滨腾越大厦两座甲级写字楼，傲然矗立在繁华的珠江新城，率先引入室中花园和自然通风采光等可持续建筑概念，得到同行瞩目。

不久，又听闻黄劲已成为广州市设计院原创最多的设计师，他多次放弃升迁做行政工作的机会，坚持在设计的第一线，每个项目都亲力亲为，亲自勾画草图，亲自建模调整体形比例。

接到黄劲厚厚的书稿时，方知其工程之多、类型之广，共有20多个项目获得建筑部和省市级奖项。其项目覆盖全国各地，并成为为数不多走出国门到东南亚做项目的建筑师。黄劲每个作品都让人耳目一新，个性鲜明。其后期作品注重南方地域性，具文化底蕴，并融入生态建筑特点。在设计手法上，擅长采用方、圆组合，空间有通透感，外形有雕塑感，已逐步形成其独特的设计风格。

随着改革开放的深入，我国建筑设计水平在不断提高，国内建筑师屡次在国际大赛中标。我们鼓励大家立足自主创新，走向世界。黄劲正当壮年，处在建筑师成熟期的黄金年代，寄望能做出更优秀作品，创作之路走得更远、更广。

2014.7.

It is the 30th Graduation Anniversary of the graduates of 1984 of South China University of Technology in 2014, and it happens the student Li Feng will graduate with the doctorate degree. He talked about his senior apprentices, that they have made their alma mater really proud by making good achievements on architectural creation and became outstanding among the young and middle-aged architects. Li Feng also mentioned Huang Jin being one of the most excellent architects from South China University of Technology who has been working at Guangzhou Design Institute for over 20 years.

Graduates of 1984 are the first enrollment after the key middle school was Restore and most of them are good calibers. Huang Jin began to show his talent at the university, that his quick designs had been always displayed in the class as examples and recognized by the teachers and classmates. When all the students graduated in 1988, Deputy Chief Architect of Guangzhou Design Institute Mr. Zhong Xinquan selected three excellent graduates including Huang Jin. After that Huang Jin has been learning and working under Academician Mr. She Junnan on hotel creation and study.

Huang Jin's first project came under the spotlight as Guangzhou Book Center. Under the guidance of Master Guo Mingzhuo, Huang Jin boldly employed triangle incision which is unique in concept, novel in shape and individualized in style, whereas the layout is foursquare and practical, and his design was highly praised by the professional jurors and government leaders. This is really remarkable for a young and inexperienced architect to deliver such excellent work.

Huang Jin cooperated with Master Guo Mingzhuo 10 years later on the successful bid of international design competition of Sun Yat-sen University and Guangdong University of Foreign Studies in Guangzhou Higher Education Mega Center, that Academician Mr. Dai Peng and I were the jurors hence I was able to know better about Huang Jin. He raised the concept of utilizing natural landscape environment and connecting the water system of two universities to form the ecological cultural axis running through the campus, which is greatly appreciated by the judges' panel and leaders of the universities. From the design, it is perceptible that Huang Jin attaches much importance to regionality and culture of the project, and endeavors to infuse the characteristic of Lingnan Architecture.

Being Huang Jin's new works of recent years, two Grade-A office building—— Qin Jian Building and Jinbin Tongyue Building proudly stand tall in the prosperous Zhujiang New Town. Sustainable architecture concept is introduced such as indoor gardens and natural ventilation and daylight, drawing the attention of his peers.

Not long ago, it is heard that Huang Jin has become the designer with the most originals in Guangzhou Design Institute. He has given up several opportunities of being promoted to executive positions and persisted to work at the front line. Huang Jin does every project by himself, such as drawing the sketches, making the model and adjusting the prototype proportion etc..

With this heavy design collection of Huang Jin's in my hand, I just get to know the quantity and variety of his works. There are over twenty projects which have won the prize of Ministry of Construction and the awards of Provincial and municipal levels. Huang Jin's projects are all over China and some of them even reach Southeast Asia. Each design of them is refreshing and characteristic. Huang Jin's latter work pays much attention to southern regionality and culture connotation, integrating into ecological architecture feature. In design approach, Huang Jin is good at using the combination of square and circle, bring transparency to the space and sense of sculpture to the shape, which have gradually grown into his unique design style.

As the in-depth development of reform and opening-up of China, our construction and design quality are continuously improving, and our architects frequently win the bids in international competitions. We always encourage that architects should be creative and international. Huang Jin is in the prime of life and in an architect's golden age of maturity. I expect him to originate better design and make more achievements on the way of creation.

He Jingtang
Summer of 2014

祝贺与期望
Congratulation and Expectation

与郭明卓先生（中）和郑启皓先生在一起
Mr. Guo Mingzhuo(betwixt) and Mr. Zheng Qihao

郭明卓
Guo Mingzhuo

全国设计大师

原任广州市设计院副院长、总建筑师，现任顾问总建筑师
国家特许一级注册建筑师

广东省土木建筑学会副理事长、建筑创作学术委员会主任、广东省注册建筑师协会会长、广州市人民政府决策咨询顾问；中国建筑学会第8届、第11届理事，体育建筑专业委员会委员。

Design master of China

Former Vice Principal and Chief Architect of Guangzhou Design Institute, Current Consulting Chief Architect, National First-Class Certified Architect

Vice President of Guangdong Association of Civil Engineering, Director of Architecture Design Academic Committee, President of Guangdong Certified Architect Association, Decision Consultant of Guangzhou Municipal Government, President of the 8th and 11th Session of Architectural Society of China, Member of Sports Architecture Committee.

拿着黄劲这本厚厚的作品集，不禁为这位辛勤耕耘，成果丰硕的建筑师感到由衷的高兴。也想写点文字来表达自己的感想。

我开始认识和了解黄劲是在20世纪90年代初，当时广州举办了羊城书市，盛况空前，市政府顺应民意，决定兴建广州购书中心。市建委要求我院与珠江院各出三个方案，供专家和领导选择。当时我负责方案设计的组织和主持，由我院三个设计室各做了一个方案参选。经过专家评审和领导选择，"新仔"黄劲的方案在六个方案中脱颖而出，被定为建造方案。这个方案两翼成直角布置，似一本"张开的书"，开口部分向南，形成裙楼、中庭和主入口。设计用几何构成手法，有现代感，大气而主题鲜明，这个项目很快建成并获得成功。后来，黄劲与法国建筑师黄福生合作的几个项目，也很成功并获得不少奖项。

我与黄劲的再度合作，是在2003年我主持广州大学城第一组团的规划设计投标的时候，黄劲带领一室的团队承担了这个任务。在规划结构上，黄劲提出的利用原有山体水系打造校园的生态轴的概念我非常欣赏。规划又通过校园的建筑形式、空间布局和中轴线与生态轴的结合，来表达中山大学历史悠久，严谨治学的人文精神，在五个组团的规划中是很具特色和个性化的。通过大学城中山大学校区的规划和建筑设计，和黄劲自己的努力，他在设计水平，团队精神和组织协调能力方面都有进一步的提高，成为业内一位知名的、成熟的建筑师。

近年黄劲自己创业，获得很大成功，设计作品遍布国内外。如鼎湖海印又一城城市设计、鼎湖总统御山庄、老挝亚欧首脑峰会接待大酒店，南昌新好景大酒店、北海银滩风帆大酒店等都是近年的作品，还有一些公共建筑如惠东博物馆和太原艺术博物馆等，成果很多。在设计手法上，除了他最常用的大尺度的几何构成手法外，在造型上也开始出现了一些非线性的手法以结合环境和迎合市场的需求。在一些酒店设计上，也采用了当地的建筑元素和建筑形式，取得很好的效果。这些都充分说明黄劲是一位与时俱进的建筑师。

借此机会，祝贺黄劲作品集出版，祝他在建筑创作道路上不断探索，超越自我，取得更骄人的成绩。

于2014年夏

Holding the wonderful design collection of Huang Jin's in my hand, I am unfeignedly delighted for this diligent and productive architect, hence I would like to also write something to express some of my thoughts.

I got acquainted with Huang Jin from the 90's during the Guangzhou Book Fair. It was an unprecedentedly prosperous fair and the Municipal Government of Guangzhou decided to build the Guangzhou Book Center responding to public opinion. Municipal Bureau of Construction requested Pearl River Institute and our Institute to respectively provide three proposals for expert and leader's selection. At that time I was in charge of the design proposals, and our three studios should create one proposal each. After the evaluation by professional jurors and leaders, the design by "freshman" Huang Jin outstood from all six designs and was appointed as final. The two wings of the architecture is deployed at a right angle like an open book, the opening facing south constitutes the podium, atrium and main entrance. The geometric approach is modern, elegant and iconic, and the project was soon built with great success. Later Huang Jing collaborated with French architect Mr. Huang Fusheng on several projects, which also made a hit and won a number of awards.

I worked with Huang Jin again in 2003 when I was in charge of the planning and design bid of the first group of Guangzhou Higher Education Mega Center, and Huang Jin led a team to take the responsibility. With regard to the planning structure, Huang Jin proposed to utilize the original water system of the hills to create the ecological axis on the campus, which I was very much appreciated. By means of the combination of campus architecture form, space layout, central axis and ecological axis to reveal the long history and academic atmosphere of rigorous scholarship, his design is very characteristic and individualized among all five groups. Through the planning and architecture design of Sun Yat-sen University of Guangzhou Higher Education Mega Center, as well as his continuous effort, Huang Jin has further improved in design quality, team spirit, organization and coordination skills, and has become a mature architect renowned in the industry.

In recent years, Huang Jin has established his own studio and gained great success. His design can be found in many places in China and overseas, the latest and fruitful projects include Dinghu Highsun Another City, Ding Hu Presidential Royal Villas, Laos Asia-Europe Summit Hotel, Nanchang Xinhaojing Hotel, Beihai Silver Beach & Sail Hotel etc., there are also a few public architecture such as Huidong Museum and Taiyuan Art Museum. Besides his skilful geometric structure in big scale, Huang Jing also adopts some non-linear approach to integrate the environment to meet the market's requirement. He also uses indigenous architecture elements and forms for specific hotel design and achieved best results. All these well proved that Huang Jin is an architect abreast of modern developments.

I would like to take this opportunity to congratulate the publication of Huang Jin's design collection. I wish he could keep exploring and surpassing himself on the way of architecture creation and become more successful.

Guo Mingzhuo
Summer of 2014

坚持设计创作的热诚
Keep Up His Passion for Design

孙礼军先生
Mr. Sun Lijun

孙礼军
Sun Lijun

现任广东省建筑设计研究院总工程师,院技术负责人
教授级高级工程师、国家一级注册建筑师

广东省优秀工程勘察设计奖评审专家
广东省土木建筑学会理事
广州市建设委员会科技委专家
广州市建设系统工程评标专家
中国建筑学会建筑师分会理事
中国建筑学会资深会员
香港建筑师学会会员
中国工程建设标准化协会常务理事
住房和城乡建设部绿色建筑评价标识专家
广东省绿色建筑评价标识专家
全国工程勘察设计行业奖评审专家

Chief Engineer and Technical Director of the Architectural Design and Research Institute of Guangdong Province, Professorship Senior Engineer, National First-Class Certified Architect

Award's Evaluation Expert of Excellent Engineering and Survey Design of Guangdong Province, Director of the Civil Engineermg and Architectural Society of Guangdong,Expert of Architecture Science and Technology Committee of Guangzhou, Expert Evaluation of Construction System Projects of Guangzhou City,Director of Architects Branch of the Architectural Society of China,Senior Member of the Architectural Society of China,Member of the Hong Kong Institute of Architects,Managing Director of China Association for Engineering Construction Standardization,Expert of Chinese Green Building Evaluation Label of Ministry of Housing and Urban-rural Development,Expert of Chinese Green Building Evaluation Label of Guangdong Province, Award's Evaluation Expert of National Engineering Survey and Design Industry

黄劲建筑师毕业于1988年,至今从业26年。我院陈朝阳院长是他的同班同学,曾多次与我提及黄劲现在仍坚持在设计院的一线工作,毕业20多年仍亲自创作,实属难得可贵,可见他对建筑设计的热爱。

与黄劲深入接触是广州大学城建设阶段,当时由黄劲主笔设计的中山大学和广州外语外贸大学项目,在国际竞赛中战胜境内外设计单位夺标。其方案巧妙地利用水系将两所大学联系于一体,将岭南园林自然地融入到校园环境当中,得到评委高度赞赏。随后我和黄劲分别担任各自项目的设计总负责人,每周一起参加大学城建设的设计例会,多次接触后,我对黄劲也有了更加深入的了解。

黄劲建筑师有较高的专业水平和协调能力,且勤奋认真,事无巨细亲力亲为,经常亲自下到工地去协调和解决设计问题,确保中山大学如期优质的建成。翻阅黄劲的作品集可以看出,黄劲个人原创作品颇多,其中以五星级酒店、教育建筑、文化建筑作品最为出色,曾获得10多项部级优秀奖,获省市奖项也就更多了。他的设计在空间上注重体现岭南建筑的文化特色,在外形上注重体现雕塑美感,已形成极其独特的个人风格。

在中青年建筑师中,黄劲已取得了相当优异的成绩,但每次与他交谈,谈及其作品时,他总认为自己的作品中精品不多,未达到其期望,他对工作精益求精,足见其对设计的热爱。行内常说,45~60岁才是建筑师的成熟期,他这个年纪已有这番成就实属难得。希望他能够秉承现在对设计的热爱,在未来的设计生涯中更上一层楼,多出些优秀的作品。

2014年7月

Huang Jin has been working as an architect after his graduation in 1988. Dean of our Institution Mr. Chen Chaoyang is Huang Jin's classmate, he mentioned sometimes that Huang Jing now still insists to work at the front line in Guangzhou Design Institution. It is very praiseworthy to keep designing by himself after over 20 years of graduation, that also fully shows his passion on architecture design.

I was able to know Huang Jin better during the construction stage of Guangzhou Higher Education Mega Center. At that time Huang Jin was responsible for the design of Sun Yat-sen University and Guangdong University of Foreign Studies, he competed with local and foreign designers and won the bid in the international competition. The design skillfully connects two universities with water system to form the ecological cultural axis running through the campus, and naturally integrates Lingnan Garden into campus environment, which is greatly appreciated by the judges. Later Huang Jin and I were in charge of respective project design, and we attended the regular design meeting of Guangzhou Higher Education Mega Center construction every Monday. I hence had more in-depth understanding of Huang Jin after times of interaction.

Huang Jin has high professional proficiency and coordination capability. Being very diligent and hardworking, he does everything by himself in all matters. He always goes to the construction site to coordinate and solve design issues to ensure the project of Sun Yat-sen University to be timely completed in good quality. From Huang Jin's design collection we can find a lot of his personal original work, among which the most outstanding projects are five-star hotel, educational architecture and cultural architecture that have won over 10 excellence awards of Department level and even more of Provincial and municipal levels. The unique personal design style of Huang Jin has come into being by attaching much importance to presenting the cultural feature of Lingnan architecture in space, and showing the sculptural beauty in shape.

Huang Jin has made outstanding achievement among the young and middle-aged architects. However, every time when I talked with him about his works, he always thinks it has not met his expectation and not enough refined products yet. Huang Jin is very enthusiastic for design as he always strives for excellence. There is a saying in our industry that the maturity period of an architect is at the age of 45 to 60. It is remarkable for Huang Jin to achieve such accomplishment at this age. I wish Huang Jin could keep up his passion for design, make more progress in his future career and originate more excellent work.

Sun Lijun
July 2014

筑梦天下——黄劲
A Dream of Architecture: Huang Jin

与王河先生在一起
Mr. Wanghe

王河
Wang He

华南理工大学博士

现任广州大学建筑设计研究院副院长、建筑总工程师、澳门城市大学国际旅游与管理学院博士导师

英国皇家特许建造师、高级建筑师、高级环境艺术设计师、硕士研究生导师

Doctor of South China University of Technology

Vice President of Architectural Design and Research Institute of Guangzhou University,Chief Engineer of Architectural Design,Doctoral Tutor of International Tourism and Management Institute of City University of Macau

Chartered Builder of CIOB,Senior Architect,Senior Environmental Art Designer,Graduate student tutor

提起黄劲建筑师，自然想到广州购书中心，这是目前广州日流量达10万人的文化建筑，建成已经有20年了，还没有哪栋文化建筑物接待人流量超过购书中心的。应该说，这是很好地解决了文化项目的城市文化、商业文化的一个典型范例。

广州大学城之中山大学和广东外贸外语大学组团，也是大学城浓沫重彩的一笔，很好地诠释了山水校园、岭南神韵的主题。从总体规划到图书馆、行政会议中心、法学院、行政学院、传播设计学院、工学院、药学院、生命科学实验大楼，能很好地把岭南建筑的庭院空间和现代高端学府糅合起来，传承了百年中山大学校区建筑所追求的深厚的岭南建筑文化底蕴的精髓。

珠江新城的CBD区的建筑是广州新城的标识。所有建筑师都希望在这里一较高低，争奇斗艳。但我在这里看到黄劲建筑师的作品却是以经济、实用、生态环保的理念很好地诠释了新岭南建筑。例如：津滨·腾越大厦、勤建大厦、云来堡酒店及办公综合楼这三个项目。我特别喜欢津滨·腾越大厦，它位于广州中轴线东侧，是一座智慧型、生态型、节能型现代化的办公大楼。采用了新产品、新技术、新空间的观念，以五层为一个绿化平台，很好地创造了微气候空间；同时也是低材高用很好的案例：以普通窗的造价设计出玻璃幕墙的效果，很好地诠释了实用、经济、美观的建筑设计法则。

酒店建筑更是黄劲建筑师的得意之作：20世纪80年代在珠三角很出名的佛山枫丹白鹭酒店、三亚福朋喜来登酒店。这两年在老挝设计的万象广晟酒店，老挝亚欧首脑峰会接待大酒店更是黄劲建筑师走向国际轨迹的见证。

我记得2012年我出版的《岭南建筑学派》一书，由清华大学两院院士吴良镛先生书写书名、著名建筑评论家曾昭奋帮忙三次审稿，书中写了近百名近现代岭南建筑师，我认为黄劲建筑师就是其中一位近30年以来，作品获奖数量最多、作品涉足领域最广、最勤奋的建筑师。

看完书稿以后，我想到了一句话："与有肝胆人共事，从无字句中读书。"天道酬勤，让我们一起领悟岭南建筑文化的博大精深吧。

2014年6月13日于桂花岗东一号

When speaking of Huang Jin, Guangzhou Book Center naturally comes to our mind, which is the cultural building with visitor flow of over 100,000 people per day. After 20 years of completion, there are no other similar buildings with more visits than Guangzhou Book Center. I must say that this is a very typical example of well resolving the urban and business culture for a cultural project.

The design of Sun Yat-sen University and Guangdong University of Foreign Studies is extraordinary notable in Guangzhou Higher Education Mega Center. It nicely interprets the theme of landscape campus and Lingnan charm. The courtyard space of Lingnan architecture and modern high-end university are perfectly incorporated through master planning, library, Administration & Conference Center, School of Law, School of Administration, School of Communications & Design, School of Polytechnics,School of Pharmacy and Life Sciences Laboratory Building,which inherits the profound quintessence of cultural connotation of Lingnan Architecture being pursued for the campus buildings of Sun Yat-sen University for over hundred years.

The buildings in the CBD of Zhujiang New Town are the icons of Guangzhou new city, and all architects would show their ultimate talent to compete in this district. Whereas I found Huang Jin's projects cost-effective, practical, ecological and environmental friendly here resulting in wonderful presentation of Neo-Lingnan Architecture, such as Jinbin·Tengyue Building, Qinjian Building, and Office Complex of Vanburgh Hotel Guangzhou, among which I am most fond of Jinbin·Tengyue Building situated on the east of the central axis of Guangzhou. This is an intelligent, ecological and energy-saving modern office building designed with the concept of new product, new technology and new space. The green platforms on every five stories create favorable microclimate environment. It also contributes to case study of clever usage of cheap materials for expensive look: the visual effect of glass curtain wall is brought out at the budget of ordinary window, giving brilliant explanation to practical, economical and beautiful design principle.

Hotel architecture is especially the strength of Huang Jin, including the very famous Fontainebleau Hotel of Foshan and Four Points By Sheraton Sanya in the Pearl Delta in the 80's. Moreover in these two years, Vientiane Guangsheng Hote and Asia-Europe Summit Hotel have witnessed the international track of Huang Jin.

As far as I remember, my design collection Lingnan Architectural School published in 2012 was name-written by double academician of Tsinghua University Mr. Wu Liangyong, and thrice examined the manuscript by renowned architecture critic Mr. Zeng Zhaofen. The book has included near one hundred modern and contemporary Lingnan architects. I believe Huang Jin is one of the architects who have won most awards, involved in most design fields, and worked the hardest.

After reading the manuscript of A Dream of Architecture —— Huang Jin, it reminded me the poem "Working with courageous people, learning from every possibilities". God helps those who help themselves, together let us perceive the extensive and profound value of Lingnan Architecture Culture!

Wang he
June 13th 2014, at No. 1 of Guihuagang East

高度的专业水平和敬业精神
Highly Professionalism and Devotion

向老挝总理、侨领姚宾先生汇报方案
Deputy Prime Minister Thongloun and Mr. Yao Bin

ລະດູໃບໄມ້ຫຼິ່ນປີ 2012, ສາທາລະນະລັດ ປະຊາທິປະໄຕ ປະຊາຊົນລາວ ໄດ້ຮັບກຽດໃຫ້ຈັດກອງປະຊຸມສຸດຍອດຜູ້ນຳອາຊີ-ເອີຣົບ, ໂດຍມີ 51 ປະເທດທີ່ມາຈາກທະວີບອາຊີ ແລະ ເອີຣົບໄດ້ສົ່ງຄະນະຜູ້ແທນເຂົ້າຮ່ວມກອງປະຊຸມຄັ້ງນີ້, ເພື່ອຕ້ອນຮັບບັນດາແຂກທີ່ເດີນທາງມາຈາກດິນແດນອັນຍາວໄກ, ພວກເຮົາໄດ້ສ້າງຕັ້ງໂຮງແຮມ ແລນມາກ ແມ່ຂອງ ລິເວີຊາຍ ຂຶ້ນຢູ່ທີ່ບໍລິເວນທີ່ມີທິວທັດອັນສວຍສົດງົດງາມແຄມແມ່ນ້ຳຂອງ.

ເພື່ອໃຫ້ໄດ້ຕາມມາດຕະຖານສາກົນ ແລະ ສະແດງອອກເຖິງວັດທະນະທຳອັນເປັນເອກະລັກຂອງລາວເຮົາ, ພວກເຮົາໄດ້ເຊີນສະຖາປະນິກທີ່ມີຊື່ສຽງຂອງຈີນ ສາດສະດາຈານ ຮວງຈິນ(HUANG JIN) ເພື່ອຮັບຜິດຊອບການອອກແບບໂຄງການນີ້. ໃນໄລຍະການອອກແບບ, ທ່ານ ຮວງຈິນ ໄດ້ໃຫ້ແບບຢ່າງສະຖາປັດທີ່ສວຍງາມຫຼາຍແບບດ້ວຍກັນ ເພື່ອໃຫ້ພິຈາລະນາຄັດເລືອກ, ພາຍຫຼັງຜ່ານການປຶກສາຫາລືກັນແລ້ວສຸດທ້າຍກໍໄດ້ຕົກລົງຄັດເລືອກເອົາຕາມແບບທີ່ຕ້ອງການ. ຫຼັງຈາກກຳນົດແບບຮຽບຮ້ອຍແລ້ວ, ສາດສະດາຈານ ຮວງຈິນ ກໍໄດ້ຮັບເຕົ້າໂຮມທີມງານອອກແບບທີ່ມີຄວາມສາມາດສູງເພື່ອລົງມືອອກແບບ ແລະ ເຮັດໃຫ້ການອອກແບບສຳເລັດໂດຍໄວ. ໃນໄລຍະການກໍ່ສ້າງໂຄງການ, ກໍ່ມີຫຼາຍຕໍ່ຫຼາຍຄັ້ງໄດ້ພົບເຫັນ ສາດສະດາຈານ ຮວງຈິນ ກຳລັງລົງຂີ້ນຳຢູ່ການອອກແບບຢູ່ສະຫນາມ, ລະດັບຄວາມຮູ້ຄວາມສາມາດອັນເລີດລ້ຳ ແລະ ຄວາມເປັນມືອາຊີບຂອງທ່ານ ໄດ້ສ້າງຄວາມປະທັບໃຈຢ່າງເລິກເຊິ່ງໃຫ້ແກ່ທຸກໆຄົນ.

ປັດຈຸບັນນີ້ ຜົນງານຂອງທ່ານ ຮວງຈິນ, ບັນກໍ່ຄືໂຮງແຮມທ້າດຄາວທີ່ທັນສະໄໝ ຕັ້ງຢູ່ແຄມແມ່ນ້ຳຂອງ ເຊິ່ງມີລັກສະນະຂອງສະຖາປັດຍະກຳລາວ ແລະ ທິວທັດສວນດອກໄມ້ອັນສວຍສົດງົດງາມ, ພ້ອມທັງໄດ້ສຳເລັດພາລະກິດ ໃນການຮັບຕ້ອນ ກອງປະຊຸມ"ສຸດຍອດຜູ້ນຳອາຊີ-ເອີຣົບ" ແລະ ອາຄານທີ່ສວຍງາມຫຼັງນີ້ ມີສ່ວນຮ່ວມປະສົມປະສານກັນຢ່າງລົງຕົວ ໄດ້ກາຍເປັນສັນຍາລັກແຫ່ງໃໝ່ຂອງນະຄອນຫຼວງວຽງຈັນ.

Landmark Mekong Riverside Hotel
ທ່ານ ຍາວບິນ

2012年秋天，老挝举办了亚欧首脑峰会，来自亚洲和欧洲的51个国家及地区派代表团前来参加，为了迎接各远道而来的贵宾，我们在美丽的湄公河畔兴建了LANDMARK酒店承担接待任务。

为了能和国际接轨，也为了能展现老挝特色文化，我们特意请来了中国著名的建筑师——黄劲教授负责本项目设计。设计期间，黄劲先生提供了多个造型优美的方案以供评审，方案几经探讨终于敲定。确定方案后，黄劲教授马上组织精干的设计队伍进行设计，快速地完成了项目的设计工作。在项目施工期间，曾多次在施工现场看到黄劲教授在指导设计工作，其高度的专业水平和敬业精神给大家留下了深刻的印象。

现在黄劲先生的佳作，一座具有老挝建筑特点和优雅园林环境的现代化五星级酒店矗立在湄公河畔，圆满地完成了峰会的接待任务，并以其高低错落的优美造型成为万象新标志性建筑。

Landmark Meckong Riverside Hotel
姚宾先生

The Asia-Europe Summit was held in Laos in the autumn of 2012. Delegations from 51 countries from Asia and Europe came far away to participate this event. We have built the Landmark Meckong Riverside Hotel by the beautiful riverside of Mekong to welcome these distinguished guests travelling a great distance.

In order to gear to the international conventions as well as to showcase the special culture of Laos, We had invited Dr. Huang Jin, a famous architect from China, to take charge of this project. During the design stage, Mr. Huang Jin has provided a multitude of proposals of graceful architecture forms for evaluation. The design was finalized after times of discussion, Then Mr. Huang Jin immediately organized his elite team to start working and efficiently completed the design. Mr. Huang Jin was frequently seen supervising at the site during construction, everyone was deeply impressed by his highly professionalism and devotion.

Today, the masterpiece of Mr. Huang's —— a modern five-star hotel integrated with Lao architecture style and elegant garden environment is standing by the Mekong River, has perfectly accomplished the mission of Summit reception, and now further becomes the new iconic architecture of Vientiane with its exquisite form of well-arranged scattered levels.

Landmark Meckong Riverside Hotel
Mr. Yao Bin

目录
CONTENTS

006	作者自序 FOREWORDS	068	勤建大厦 Qinjian Building
008	业界与客户的评价 EVALUATION FROM PREES AND CUSTOMERS	074	广州购书中心及维多利广场 Guangzhou Book Center & Victory Plaza
020	前言 PREFACE	084	广州大学城购书中心 Book Center of Guangzhou Higher Education Mega Center
032	**城市设计** **URBAN DESIGN**	088	新华大厦 Xinhua Building
034	广东省肇庆新城市CBD核心区规划（鼎湖海印又一城） New Downtown CBD Nodal Region of Zhaoqing of Guangdong (Ding Hu Haiyin Youyicheng)	096	佛山市世博商业中心 World Expo Business Center of Foshan
040	广州大学城组团一的规划设计（中山大学和广东外语外贸大学） The First Group Participation-Guangzhou Higher Education Mega Center (Sun Yat-sen University & Guangdong University of Foreign Studies)	102	佛山市南海区国土局地籍资料库 Cadastral Database of Land and Resources Bureau of Nanhai District of Foshan
		108	松岗广场 Songgang Square
048	广东省连山小三江温泉度假区 Xiaosanjiang Hot Spring Resort of Lianshan of Guangdong	114	佛山市南海区国家税务局综合业务用房 Integrated Service Office of State Administration of Taxation of Nanhai District of Foshan
050	广东省连平县南湖蝴蝶谷温泉度假区 Nanhu Butterfly Valley Hot Springs Resort of Lianping of Guangdong	120	南宁市云星办公楼 Yunxing Office Building of Nanning
052	广州市增城小楼慢城 Xiaolou Slow City of Zengcheng of Guangzhou	122	惠州市惠东博物馆 Huidong County Museum of Huizhou
054	**公共建筑** **PUBLIC BUILDINGS**	**124**	**酒店建筑** **HOSPITALITY**
058	津滨·腾越大厦 Jinbin · Tengyue Mansion	126	北海市银滩风帆大酒店 Silver Beach & Sail Hotel, Beihai
		130	海南省三亚市福朋喜来登大酒店 Four Points by Sheraton Sanya, Hainan Province

142	老挝亚欧首脑峰会接待大酒店 Landmark Meckong Riverside Hotel Landmark Meckong Riverside Hotel of Asia-Europe Meeting Summit Hotel, Laos		Campus, Sun Yat-sen University
156	老挝万象广晟大酒店 Guangsheng Hotel, Vientiane, Laos	208	中山大学东校区工学院 School of Polytechnics of East Campus, Sun Yat-sen University
160	广州云来斯堡酒店、办公综合楼 Vanburgh Hotel, Office Building	212	中山大学东校区微纳尺度材料及生命科学实验大楼 Nanoscale Materials & Life Sciences Laboratory Building of East Campus, Sun Yat-sen University
176	佛山市南国桃园枫丹白鹭酒店 Fontainebleau Hotel, Nanguo Peach Garden, Foshan	220	中山大学东校区行政会议中心 Administration & Conference Center of East Campus, Sun Yat-sen University
182	鼎湖·总统御山莊 Ding Hu·Presidential Royal Villas	226	中山大学东校区工科实验楼与药学院楼 Engineering Laboratory Building & Pharmaceutical Building of Sun Yat-sen University, Guangzhou Higher Education Mega Center
188	**教育建筑** **EDUCATIONAL BUILDINGS**		
190	中山大学东校区图书馆 Library of the East Campus, Sun Yat-sen University	230	广州科技职业技术学院 Guangzhou Vocational College of Science and Technology
196	中山大学东校区公共教学楼 Public Classroom Building of East Campus, Sun Yat-sen University	236	广东财经大学华商学院 Huashang College of Guangdong University of Finance & Economics
198	中山大学东校区法学院 School of Law of East Campus, Sun Yat-sen University	242	南昌市安义教育园 Education Park of Anyi County, Nanchang
200	中山大学东校区行政学院 School of Administration of East Campus, Sun Yat-sen University	248	桂林航天工业学院来宾校区 Laibin Campus of Guilin University of Aerospace Technology
204	中山大学东校区传播与设计学院 School of Communications and Design of East Campus, Sun Yat-sen University	**252**	**住宅** **RESIDENTIAL**
206	中山大学东校区基础实验楼 Fundamental Laboratory Building of East	**255**	**后记** **POSTSCRIPT**

前言
PREFACE

与执信中学的恩师和学友一起
Zhixin High School Metors and Alumni

2012年起，受广州大学之邀，我每周抽出两天时间到建筑系为学生授课，希望能将多年的实践经验传授给下一代建筑设计师，让他们能在社会上做出优秀的建筑作品。以此回馈社会，尽一个公民的社会责任。

在授课期间，总感觉提供给学习的课件资料还不够直观充实，还没能把我20多年的建筑设计成果真实地呈现出来，而我的导师谭卓枝先生也多次建议我出版一本个人设计作品集，可以更系统地总结过往的经验，更好地为今后的设计工作进行规划。因日常的设计任务繁忙故一直未能进行，直至在AEC建筑+设计沙龙联展与论坛上遇到策展人/出版人叶飚先生，其协助我重新整理相关的项目资料，拍摄已经落成项目的实景照片，让我能更轻松地实现这出版的愿望。同时，我儿黄思宇也参与到本书繁重的资料整理和编辑工作中，他自入读大学后一直协助我的设计工作，特别在草图阶段建模做了大量工作，并在方案比选时提出了很多很好的意见，如老挝广晟酒店、广东科技职业学院、老挝商业城、惠东县博物馆前期等项目设计工作。黄思宇从小酷爱画画，曾获全国少儿绘画大赛二等奖一项，三等奖四项。最让我欣慰的是，其在高中阶段学习排名从入学时350名升至毕业时前15名，并闯入重点本科学习建筑学，其大学期间在重点班的设计成绩均名列前茅，代表学校参加广东省五校建筑系并获奖。

而我从事的建筑设计工作也是与儿时的喜好甚有关系的，喜欢画画的我在小学就曾临摹了多本《三国演义》的"小人书"，但因在郊区小学就读未能接受更专业系统的绘

At the invitation of Guangzhou University, I give lessons to the students of Architecture Department at Guangzhou University twice a week from 2012, in the hope of nurturing the architectural designer of next generation with my hands-on experience, and developing their ability to create excellent architecture in the future, as my reward to the society and obligation as a citizen.

During the classes, I find that the learning materials are not direct and practical enough to truly manifest my architectural design in the past 20 years, and my mentor Mr. Tan Zhuozhi always suggests me publishing personal design collection, this could systematically summarize my past experience, and plan my future design work better. However, it has never been implemented due to my busy schedule, until I met exhibition planner / publisher Mr. Ye Biao (Billy Yip) on the AEC Architecture + Design Salon Joint Exhibition & Forum, who later assisted me to organize the pertinent project materials and take photos of the accomplished projects, thus I am able to realize the wish in an easier way. Meanwhile, my son Huang Siyu also participated in the material collection and editing of this Book, who has been joining my design projects since his university study. He has done a good quantity of work especially in sketching and modeling stage, and given many good ideas in scheme comparison, and involved in projects such as Guangsheng Hotel in Laos, Guangzhou Vocational College of Science and Technology, Laoyue Commercial City in Laos, Huidong County Museum etc. Siyu loves drawing from his childhood and has won one second prize and four third prizes in National Children's Drawing Contests. It is a gratification to me that his ranking raised from No.350 in enrollment to No.15 when graduated from senior high school. He then studied in the Department of Architecture at a key university and came out top in the key class. Moreover, on behalf of the University, he participated in the Quick Design Competition held by the Department of Architecture of five

儿时绘画素材
Picture-story Booklets of Romance of the Three Kingdoms

母校——广州执信中学
Zhixin High School

母校——华南理工大学
South China University of Technology

恩师——佘畯南院士
Mr. She Junnan

与郭明卓大师一起
Mr. Guo Mingzhuo

画训练,稍大每当假期有机会,我在少年宫绘画班的窗外偷偷看老师教别人绘画。后来就读广州执信中学,那里的建筑是具浓郁岭南建筑文化的气息,红墙绿瓦的风格与华南理工大学的红楼建筑甚相似,这对我日后申报华南理工大学的建筑专业有极大的影响,1988年我在蔡婉班主任的指引下,如愿入读华南理工大学建筑学系。

在华南理工大学我受到了众多有名教授的系统教导,有叶荣贵、刘烨先生等教授基础,黎显瑞、谭伯兰、赵伯仁先生等教授专业课,张锡麟系主任带我所在的毕业班,以及来自同济大学的龚耕先生为我们讲解设计方法论,华南理工大学建筑设计院的何镜堂院士等先生更经常过来指导,为我们打下了较扎实的专业基础。期间,我利用每届暑假的时间在各地写生,如深圳、上海写生现代建筑,北京、苏州、无锡写生古典园林,将近500张的速写画提升了我对建筑设计的构思和表达能力,更加深了对中国古典建筑的认识与热情。

毕业后进入广州市设计院,有幸师从佘畯南院士、钟新权先生多年进行建筑设计,跟随他们到白天鹅宾馆、汕头金海湾酒店等项目现场,聆听院士讲解设计构思,使我从思维到操作上重新理解和认识建筑设计的原则。佘畯南院士的教导至今铭刻于心,其中"做设计首先要学会做人"和"建筑是研究人的艺术",更指导我去完成每一个设计。

我的第一个建筑设计作品是广州购书中心,是当时广州市黎子流市长亲抓的重点项目。设计在郭明卓大师、陈树棠先生的指导下

universities in Guangdong and honorably won the prize.

Architectural design I dedicated to is quite relevant with my childhood hobby. I used to copy a couple of picture-story booklets of Romance of the Three Kingdoms, however, I could not get more professional drawing training because I studied in a suburban primary school. When I grew older, I always peeked into the windows of the drawing class of children's palace and learn by myself. Later I studied in Zhixin High School, where the buildings are full of distinct architectural culture of Ningnan style, the red walls and green tiles is similar to the Red House Architecture of South China University of Technology, which greatly influenced my decision to apply for Architecture major of this University. In 1988, finally I achieved my goal of studying in the Architecture Department of South China University of Technology under the guidance of Class Tutor Cai Wan.

I was honored to be taught by many famous professors in South China University of Technology thus established solid foundation, such as Mr. Ye Ronggui and Mr. Liuye for fundamental classes; Mr. Li Xianrui, Mr. Tan Bolan and Mr. Zhao Boren for professional classes; Mr. Zhang Xilin, Director of Department for instructing our graduate class; Mr. Gong Geng from Tongji University for teaching design methodology; Academician Mr. He Jingtang for frequent guidance. I travelled to different cities to draw from life during summer vacations, for example, I went to Shenzhen and Shanghai to draw modern architecture, to Beijing, Suzhou, Wuxi to draw traditional gardens. Around 500 sketches enhanced my conception and expression towards architectural design, and reinforced my understanding and passion on Chinese traditional architecture.

I started my career in Guangzhou Design Institute after graduation and worked under Academician Mr. She Junnan and Mr. Zhong Xinquan on architecture design for several years. I followed them to visit the project venues such as the White Swan Hotel in Guangzhou and Shantou Golden Gulf Hotel etc. By learning design concept from the seniors, I reunderstood and reperceived the principle of architectural design from ideation to operation. I still bear in mind the edification from Mr. She Junnan, that "Learn to live is the foundation of design" and "Architecture is the art of studying people" have always been the guideline of my design.

与恩师——钟新权先生（中）在一起
Mr. Zhong Xinquan(betwixt)

与中山大学建筑设计组同事在一起
Project Team(Sun Yat-sen University)

与马震聪先生和郑启皓先生在一起
Mr. Ma Zhencong and Mr. Zheng Qihao

与法国著名建筑师黄福生先生在一起
Mr. Huang Fusheng

进行，初出道的我日夜构思，最终确定了最优方案：根据群体建筑的45°轴线，自然切割广州购书中心所在地块，生成三角形组合的形体。从项目周边环境考虑，寻找设计构思灵感，成为我日后的重点方式。该项目建成后获得全国优秀工程设计铜质奖、住房和城乡建设部级优秀设计二等奖、广东省及广州市优秀工程设计一等奖。而18年后我设计的广州第二个大型购书中心——新华大厦于2011年金秋落成营业了，本项目用地非常窄小，且面对城市重要道路交会处，我在设计上将建筑形体解构为大折板和弧面，立面犹如打开的手提电脑和张开的书本，隐喻现代书城的特点，门框式造型和弧面朝向，顺应了路口转弯角的景观和形象要求，外形追求强烈雕塑美感，较完美地解决了空间与形式的关系，建筑形体有变化但平面又具实用性。

我的第二个设计作品是中山市第一座超高层甲级写字楼——中国银行大厦，项目是在谭卓枝先生带领下完成的，我在其顶部建筑设计中融入了中国古代钱币的图案构成。在钟新权先生的指导下，我的第一个超五星级酒店项目设计经赛中标了，佛山市兆银大酒店开始实施后更得到佘院士提供的许多宝贵意见，酒店除拥有空间层次丰富的绿化中庭，每五层的客房楼层更共享一个空中花园，这是我首次将岭南建筑元素运用于高层建筑之中。从本项目开始关小梅女士长期为我深化建筑设计，后来多个成功项目均有其辛勤的劳动。

来自法国的黄福生先生是我很尊敬的国际著名建筑师，其高度的敬业精神，重视融

My first design is Guangzhou Book Center, a key project in the charge of Mr. Li Ziliu, Mayor of Guangzhou. Under the instruction of Mr. Guo Mingzhuo and Mr. Chen Shutang, I worked day and night and finally confirmed the optimal solution: the land of Guangzhou Book Center is naturally incised by the axis at the angle of 45 degrees of the building complex forming the triangle shape. The similar practice of seeking inspiration from the peripheral environment has become the keystone of my future design. This project has later obtained Bronze Award of State Excellent Design, Second Prize of Excellent Design Award by Ministry of Housing and Urban-Rural Development, and First Prize of Excellent Engineering Design of Guangzhou and Guangdong. 18 years later, my second large book center project in Guangzhou—Xinhua Building was completed and came into service in autumn of 2011, which is set on the limited land facing an important intersection in downtown. The architectural form is deconstructed in large folded plate and curvy surface, and the elevation is like an open laptop and unfolded book, which implies the feature of modern book center. The door frame shape and orientation of curvy surface is in compliance with the landscape and image at the corner, ensuring strong sculptural beauty, which ideally solve the relationship between space and form, resulting in changeable architectural shape and practical plane.

My second project is Bank of China Building—the first ultra high-rise Grade-A office building in Zhongshan. This project is led by Mr. Tan Zhuozhi and I infused the pattern of Chinese ancient cooper coin into the design of the architecture top. My first super five-star hotel Zhaoyin Hotel of Foshan won the bid with the help of Mr. Zhong Xinquan and gained valuable opinions from Academician Mr. She. Besides the green atrium of rich space layers, every five guestroom stories of the hotel share one hanging garden, which come first to me to utilize Lingnan architectural elements to high-rise buildings. Ms. Guan Xiaomei commenced to polish the architectural design for me from this project, who keeps contributing to many of our successful projects since then.

Mr. Huang Fusheng from France is one of the famous international architects who I respect most. He is highly professional and pays lots of attention to integrate the original history and culture into the project from whom I benefit much. Mr. Huang focuses on the project concept and I deal with the construction documents. We worked harmonious together to accomplish five projects including Fontainebleau

与设计师们在一起
Team of Designers

向老挝总理、侨领姚宾先生汇报方案
Deputy Prime Minister Thongloun and Mr. Yao Bin

与设计师们在一起
Team of Designers

合项目地域的历史、文化的设计观念让我受益良多。由黄先生负责方案构思，我方负责深化施工图设计，共同合作完成的佛山市南国桃园枫丹白鹭酒店、枫丹白鹭酒店国际会议中心、南海区国土局地籍资料库、南海区国家税务局综合业务用房、松岗广场等5个项目，更获得部级、省市级等5个奖项。乃至10年后，我主笔设计老挝五星级 LANDMAR 酒店时，还特邀黄先生亲临老挝万象指导设计。

与郭明卓设计大师共事了20多年，郭大师指导我完成了广州购书中心、广州大学城中山大学等项目。其对项目设计具有敏锐的目光、独到的点题，让我十分佩服。在广州大学城中山大学方案投标过程中，为了能有一个优胜的构思方案，我和郑启皓、周茂、蔡展刚、钟献荣等建筑师组成的五人方案组日夜不停地封闭工作，第一轮方案投标的构思在凌晨3点蓄势并发的情形还历历在目。由此我坚信：方案设计是没有天才论的，只有经过深入周密的思考，进行所有可能性方案的比较之后，才会得到一个较好的构思方案，好的方案就等于20%的灵感加上80%的努力。

而后我作为广州大学城中山大学设计方总负责人驻场，家与工地是每天踏着晨光出发，披着星月归去，尽力全过程地监控整个施工进程，及时修正设计成果，从总体建筑至每个建筑细部的处理，乃至建筑材料的选择，为达至最佳效果而尽最大的努力。目睹一座座教学楼拔地而起，一棵棵幼苗茁壮成长，广州大学城中山大学的面貌日新月异地变化，心里备感振奋。当广州大学城中山大学校区一期工程落成后，其以浓厚的校园文化及岭

Hotel of Nanguo Peach Garden in Foshan, International Convention Center of Fontainebleau Hotel, Cadastral Database of Land and Resources Bureau of Nanhai, Integrated Service Office of State Administration of Taxation of Nanhai District and Songgang Plaza, with which we received five awards at municipal, provincial and ministry levels. Ten years later, when I was in charge of the five-star Landmark Hotel in Laos, Mr. Huang was invited to work with us and give instruction in Vientiane of Laos.

I have been working with design master Mr. Guo Mingzhuo for over 20 years, with whose help I accomplished the projects of Guangzhou Book Center and Sun Yat-sen University in Guangzhou Higher Education Mega Center etc. I highly admire Mr. Guo for his excellent taste and acute sense towards project design. In order to work out an outstanding concept for Sun Yat-sen University in Guangzhou Higher Education Mega Center, I went all out and worked with Mr. Zheng Qihao, Mr. Zhou Mao, Mr. Cai Zhangang and Mr. Zhong Xianrong day after night during the bidding process. It still came to my mind that we finally exerted to ful fill the concept at 3 AM for the fist round of bidding. Hence I firmly believe that there is no genius in design, a better concept comes from in-depth and careful thinking, after comparison with all possibilities. A good concept is the combination of 20% inspiration plus 80% efforts.

Later I stationed at Sun Yat-sen University in Guangzhou Higher Education Mega Center as the designer-in-charge. I arrived at the venue in early morning and left in late night and tried my best to supervise the entire progress to achieve the best result, such as modification of the design in due time, manipulation from the master construction to every detail, and even selection of building materials. I was so excited seeing the erecting of buildings and growing of sapling, and the amazing changes on the campus. When the first phase was accomplished, it was highly praised for the rich campus culture and architecture of Lingnan Style. The project has also won a couple of awards including Excellent Award of Architectural Society of China Architecture Creative Awards, Excellent Survey & Design Award of Ministry of Housing and Urban-Rural Development, and First Prize of Excellent Engineering Design of Guangzhou and Guangdong. It was such a memorable experience to participate in the construction of Guangzhou Higher Education Mega Center, create good study environment for young students and benefit our next generations.

广州购书中心
GuangZhou book center

维多利广场
Victory Plaza

中山市中国银行大厦
Bank of China Tower, Zhongshan

佛山市兆银大酒店
Zhaoyin Hotel of Foshan

南建筑特色得到各方好评，更获得中国建筑学会建筑创作佳作奖、住房和城乡建设部级优秀勘察设计奖、广东省及广州市优秀工程设计一等奖等多个奖项。有幸参与广州大学城的建设，为莘莘学子提供良好的学习环境，造福子孙后代是我们每个设计人难忘的经历。

广州大学城中山大学建筑群设计是集体智慧的结晶，图书馆是郑启皓先生与我合作主笔设计的，其犹如知识宝盒和打开的书本，造型极具现代雕塑美感，其在校园中轴线上具有恢宏的尺度和象征意义；在广州大学城中山大学一期公共教学楼、行政学院、法学院、传媒学院、基础实验楼、工学院等项目设计中，我的知心好友和精诚的同事周茂、蔡展刚、钟献荣、吴树甜、林浩骏、李明、吕向红等建筑师与我在项目设计和现场服务中倾注了大量心血，同时其他专业设计师蔡伟平、王松帆、丰汉军、高蕴瑶、华锡锋、屈国伦、李明盛、朱宗明、沈耀忠等对我们建筑设计给予大力支持。

2005年，广州大学城中山大学二期的行政会议中心、微纳尺度材料及生命科学实验大楼、药学院、动物实验中心等项目启动，我在建筑设计的风格上延续了一期的特点，针对空间和立面的处理上更为精彩和细致。而新加入工作组的郑晓山、黄考颖、黄蕙青、候则林、罗铁斌等设计师在项目设计中表现出色。

在广州大学城中山大学的建筑设计实施评审过程中，何镜堂院士更三次亲自评审图纸，而赵路院长、马震聪、张南宁、陈小珉、

The design of Sun Yat-sen University in Guangzhou Higher Education Mega Center is the result of collective wisdom. The library is the work collaborated by Mr. Zheng Qihao and me, whose thesaurus box and open-book shape is full of modern sculptural beauty, and its location on the central axis of campus is of extensive scale and symbolic meaning; the Public Classroom Building, School of Administration, School of Law, School of Media and Communications, Fundamental Laboratory Building and School of Polytechnic are designed by my dear friends and hardworking colleagues Zhou Mao, Cai Zhangang, Zhong Xianrong, Wu Shutian, Lin Haojun, Li Ming, Lv Xianghong. We have put in great efforts in project design and on-site service, meanwhile other professional designers such as Cai Weiping, Wang Songfan, Feng Hanjun, Gao Yunyao, Hua Xifeng, Qu Guolun, Li Mingsheng, Zhu Zongming and Shen Yaozhong have given substantial support for our architecture design.

In the year of 2005, several projects of the second phase of Sun Yat-sen University in Guangzhou Higher Education Mega Center were initiated including Administration and Conference Center, Nanoscale Materials & College of Life Sciences Laboratory Building, Pharmaceutical Building, Animal Laboratory Building. The design style has extended the same feature of the first phase, however the manipulation of space and elevation is more delicate and detailed. And the new team members such as Zheng Xiaoshan, Huang Kaoying, Huang Huiqing, Hou Zelin and Luo Tiebin have done a great job in the project design.

During the course of project examination of Sun Yat-sen University in Guangzhou Higher Education Mega Center, Mr. He Jingtang has personally gone through the blueprint for three times, and Mr. Zhaolu, Ma Zhencong, Zhang Nanning, Chen Xiaomin, Huang Xiaolong and Mr. Huang Daren, Principal of Sun Yat-sen University also have given full support and valuable advice which benefited me a lot. The project collection is then integrally summarized in the book The Best University of South China edited and published by the publishing house of Architectural Creation of Beijing Institute of Architectural Design.

I was in charge of the planning and architecture design of Huashang College of Guangdong University of Finance & Economics and Guangzhou Vocational College of Science & Technology after that, in these two projects practicability and cost

佛山市南国桃园枫丹白鹭酒店
Fontainebleau Hotel, Nanguo peach Garden, Foshan

南海区国家税务局综合业务用房
Integrated Service Office of State Administration of Taxation of Nanhai District

佛山市南海区国土局地籍资料库
Cadastral Database of Land and Resources Bureau, Nanhai District

黄小龙先生以及中山大学黄达人校长等也一直大力支持，提出许多宝贵意见，让我受益颇丰。关于本项目的汇总在由北京市设计院《建筑创作》杂志社编辑出版的《华南第一学府——中山大学》中有很完全的总结。

后来，我在承接广东财经大学华商学院和广州职业技术学院两所大学的规划与建筑设计的任务时，十分注重针对于民营大学的实用性和经济性。而在2000年我们通过全国竞赛中标，我获得到了全国第一个县级教育园区的设计任务。对于教育建筑设计的多年经验心得，我认为需要重点体现不同的校园文化特点和氛围。

在广州大学城中山大学一期项目开学使用时，广州天河新中轴线上的一组超高层城市综合体——维多利中心亦建成并投入使用了，这是10多年前由陈树棠先生指导我完成方案和初步设计的项目，其后由广州市城建开发总公司设计院完成施工图的设计。截至2012年，我在广州新中轴线上及珠江新城CBD核心区的建筑设计作品共有5个获得了部级及省级奖，我成为广州中轴线上作品量最多的建筑师。特别是位于广州新中轴线上的津滨·腾越大厦，这是一座具有新岭南风格的智慧型、生态型办公楼，大楼利用空中花园引导穿堂风，立面采用分段式构图和自然换气系统，防水内开启窗、仿玻璃幕墙效果等绿色创新技术，让办公楼有独特形象和较高的技术含量。本项目更获得住房和城乡建设部级优秀设计二等奖，这是对我在超高层建筑上运用新岭南建筑设计思想的充分肯定。前段日子，我与学生谈及津滨·腾越大厦的三点设计体会：

effectiveness are two key concerns. We won the bid in a national competition in 2000 and gained the project of the first education parks of county level approved by the State. According to my years of experience in educational architectural design, I think it is necessary to present various cultural and environmental features of different campus.

At the time that the first phase of Sun Yat-sen University in Guangzhou Higher Education Mega Center was brought into operation, the ultra high-rise complex on the new central axis of Tianhe District of Guangzhou—Victory Plaza also came into service. This project was designed by me over ten years ago under instruction of Mr. Chen Shutang, and the construction drawing was accomplished by Design Institute of Guangzhou Urban Construction Development Corporation. Up to 2012, five of my design projects on the new central axis of Guangzhou and at the core of Zhujiang New Town CBD have obtained awards of provincial or ministry levels, thus I have become the architect who owns the most works on the central axis of Guangzhou. Hence I would like to mention Jinbin · Tengyue Building that sits on the new central axis of Guangzhou, which is an intelligent and ecological office building of neo-Lingnan style. The Building has employed hanging gardens to lead cross-ventilation. The elevation is deployed with innovative green technologies such as sectional way with natural airchanging system, waterproof inner-open window and glass curtain wall-imitating effect, enabling unique image and high-tech sense to the office building. The award of second prize of Excellent Design of Award by Ministry of Housing and Urban-Rural Development is the recognition of my neo-Lingnan architecture design for ultra high-rise buildings. A few days ago, I mentioned three points learned from the design of Jinbin · Tengyue Building: firstly is to insist on neo-Lingnan style and present the regional architecture culture; secondly is to innovate and apply modern architecture technology; thirdly is to pursue the perfect combination of space and form. I treat every project as artwork and try to achieve the fusion of space, art and technology. It is my preferred way to design the simple architecture form in square, circle and triangle and seek for the sculptural beauty of architecture.

You can always find something to learn from the people who keep you company, my professional knowledge is widened by studying from developers and management companies, such as Mr. Huang Naixian, owner of Zhongshan Cui Jing Hotel, Mr.

广州大学城中山大学一期校区
Sun Yat-sen University in Guangzhou Higher Education Mega Center

广东财经大学华商学院校区
Huashang College of Guangdong University of Finance & Economics

广州科技职业技术学院校区
Guangzhou Vocational College of Science and Technology

首先是坚持新岭南建筑风格取向，展现地域建筑文化内涵是根本；二是现代建筑技术的创新和运用；三是追求对空间与和形式的完美结合。我会把每个作品当作艺术品去对待，力求做到空间与艺术、技术的融合，偏好运用方、圆、三角等简洁的建筑体形来进行设计，追求建筑的雕塑美感。

三人行必有我师，虚心向开发商、管理公司学习更扩大了我的专业知识面，如中山翠景大酒店的业主黄乃咸先生，老挝LANDMARK酒店的于震和周景行，海印集团的邵建佳和郑铭枢先生，都向我传授了许多酒店管理方面的经验，让我更深层次地理解国际星级酒店的营造，使我能顺利承接了多间五星级酒店的设计。如已经完成的大型的餐饮与娱乐型酒店有中山市五星级翠景大酒店，海滨度假酒店有海南三亚福朋喜来登酒店（与香港巴马丹拿事务所合作设计），商务型酒店有广州五星级云来斯堡酒店，国宾馆有老挝五星级LADNMARK酒店，另外还有正在设计的北海五星银滩大酒店、广晟国际大酒店、鼎湖总统御山庄五星级大酒店和哈尔滨东方红温泉大酒店等。

我认为酒店设计除了需要处理好功能流线和熟识酒店营运管理外，重视酒店的文化在设计中的应用尤为重要，如在北海银滩大酒店和三亚福朋喜来登大酒店设计中，建筑外形以时尚的风帆造型与大海文化相呼应；在广州云来斯堡酒店的项目设计中，在现代风格中融入欧式建筑元素，以吻合塑造近代商都广州十三行风情的这一酒店经营主题；肇庆鼎湖总统御山庄则以中式风格和院落的

Yu Zhen and Mr. Zhou Jingxing of LANDMARK Hotel of Laos, Mr. Shao Jianjia and Mr. Zheng Mingshu of Highsun Group. They have imparted me the experience of hotel management, hence I was able to understand more about the creation of international star hotel and successfully designed several five-star hotels, for example: large entertainment hotel such as Cui Jing Hotel of Zhongshan (five-star); seashore resort hotel such as Four Points By Sheraton Sanya of Hainan (cooperated with Hongkong Palmer & Turner Group, P&T); business hotel such as Vanburgh Hotel of Guangzhou (five-star); State Guesthouse such as LANDMARK Hotel of Laos (five-star), and those projects at design stage such as Silver Beach & Sail Hotel of Beihai (five-star), Guang Sheng International Hotel, Ding Hu · Presidential Royal Villas and Harbin Dongfanghong Hot Spring Hotel etc.

Apart from the manipulation of functional streamline and knowledge of hotel operation and management, I believe it is crucial to attach much importance to the application of hotel culture in the design. For example, the fashionable sail-shape architecture form echoes the sea culture in Silver Beach & Sail Hotel of Beihai and Four Points By Sheraton Sanya of Hainan; modern style is infused with European architecture elements in Vanburgh Hotel of Guangzhou, to match the flavor of "Thirteen Factories" of modern Guangzhou as a business city; the combination of Chinese style and courtyard space compliments with the mountain and waterscape of Dinghu to create landscape theme of cultural connotation for Ding Hu · Presidential Royal Villas.

Through years of experience in hotel design, the most intractable and challenging project is the LANDMARK Hotel of Laos. In October 2011, I received a call from Mr. Yao Bin, an oversea Chinese in Laos, inviting me to design a brand new five-star hotel for Asia-Europe Summit 2012 attended by top leaders from 51 countries. After arduous rounds of comparison and optimization, finally the design concept of major Lao Style with a touch of European and Chinese elements was confirmed. As I reported the Government of Laos, Deputy Prime Minister Thongloun has highly praised it. It took only one year from design to delivery of LANDMARK Hotel of Laos. In order to outstand the ethical and cultural feature of Laos, I visited almost every famous architecture of Laos, I listened to the Buddhist term in many temples to absorb the profound Buddhism Culture of south-east Asia. Strong Lao architecture characteristic is manifested on the blending of modern style and Lao folk-custom on rooftop and columns on

津滨·腾越大厦
Jinbin · Tengyue Building

中山大学东校区行政会议中心
Administration & Conference Center of East Campus, Sun Yat-sen University

中山大学东校区微纳尺度材料及生命科学实验大楼
Nanoscale Materials & Life Sciences Laboratory Building of East Campus, Sun Yat-sen University

空间组合，并与鼎湖山水相映衬，为酒店塑造极有文化沉淀的山水主题。

在酒店设计的历程中，最让我感觉棘手和挑战的是2011年10月爱国侨领姚宾先生从老挝打来的电话，其邀请我为在老挝举行的2012年亚欧51国首脑峰会设计一间全新的五星级国宾馆。接受任务的我经过艰苦的多轮方案比较，最终确定了以老挝风格为主，加入欧式、中式元素的设计构思。在我的设计方案向老挝政府汇报后，得到了通伦总理的高度赞赏。老挝LANDMARK酒店从建筑设计到交付使用仅用了1年时间，而为了体现老挝的民族文化特点，我几乎遍访了老挝全境有名的建筑，深入多个寺庙聆听佛语，感受到东南亚地区深厚的佛教文化。设计体现浓厚的老挝建筑特点，在屋顶、立面柱式等用现代手法融合老挝民俗元素，基座和柱式则借鉴欧式建筑的构图特点，建筑外形恢宏大气，并吸取了中国建筑群落组合的优点，设计采用老、中、欧式混合搭配的营造法，更体现了东南亚多元文化融合的特点。文化特征是酒店的灵魂，是酒店建筑设计的出发点。在老挝项目设计中，我明显感受到随着我国经济不断向外发展，我们的文化和第三产业的国际地位也不断提高，中国建筑设计在国际上有了更重要的地位，中国建筑师应有充足的自信心和自豪感，大胆走出国门。但要留意的是，设计总承包更能满足业主要求，国外设计需要建筑师有相当高的综合素质，除了有优秀的建筑方案，更需有各专业的知识及协调的能力。

近年，我所设计的地产项目作品也遍布全国，在广东、广西、江西、山西、黑龙江等

elevation; the pattern composition on pedestals and columns is drawn from European architecture; the imposing and grand shape introduces the merit of Chinese building complex; the mixture of Lao, Chinese and European styles showcases the fusion of diversified culture in south-east Asia. Culture is the soul of a hotel, and the starting point of hotel design. From this project, I can clearly feel the outspread economy of our country and the increasing international status of Chinese culture and the third industry. Architectural design of China is playing a more important role in the world, and Chinese architects should nourish sufficient confidence and pride and jump out bravely. However, it should also be brought to attention that, in order to meet all demands of the client, the architect working on the project abroad is required to possess high comprehensive quality. Besides excellent solutions, professional knowledge and coordination capability is also a must.

The projects I designed in recent years have spread all over China, such as Guangdong, Guangxi, Jiangxi, Shanxi and Heilongjiang and so on. One of the projects, the ultra-large commercial property—World Expo Business Center of Foshan is currently under construction. My design for the project has succeeded in the bid, and the designers including Huang Junpeng and Fang Liangbing have polished and perfected the construction drawing. My major concern for commercial property is the principle that the design should be integrated with the development concept.

Urban planning is one of the aspects I pay much attention to in these years and deemed as the extension of extent and depth. My second planning design is the 230 hektare project at the new downtown CBD nodal region of Zhaoqing (Highsun Youyicheng), which abstractly integrate two cultural themes of Judge Bao's history and Dinghu landscape. The ongoing project of Harbin Dongfanghong Hot Spring Hotel has already attained approval from the Government, which is a 666-hectar hot spring resort with the theme of mountain and waterscape as well as ice-snow culture. In my point of view, for any planning conducted by an architect, the advantages of building complex and organization should be shown, and the key issues like society, economy and development concept should be attended.

Besides the seniors and colleagues mentioned above, there are many designers as well contributed to my designed projects, who should share the same honor for

中山市翠景大酒店
Cui Jing Hotel, Zhongshan

新华大厦
Xinhua Building

勤建大厦
Qinjian Building

南方铁道大厦
Southern Railway Building

省份均有众多项目。其中由我本人方案中标，由黄军鹏、方良兵等设计师深化和施工图设计的超大型商业地产——佛山世博中心正在实施过程中，我对于商业地产设计首要考虑的是设计构思需与开发理念相结合的原则。

城市规划也是我近年甚为关注的方面，它们是建筑设计在广度、深度的延伸。我第二个规划设计项目是肇庆新城市CBD核心区（海印又一城）230公顷规划，项目抽象地融入了包公历史、鼎湖山水的两个文化主题；已通过政府评审正在进行中的哈尔滨东方红项目，是一个666公顷温泉旅游区规划，以山水环境、冰雪文化为规划主题。我觉得建筑师做规划，要发挥在建筑群体空间、组织上的优势，并要关注社会、经济、开发理念等重大问题。

除文中提及的师长、同事外，还有许多设计师在项目设计中贡献了力量，这些成果也是属于您们的，在此特向以下成员表示感谢：郭明卓、谭桌枝、钟新权、董学奎、赵路、陈树棠、马震聪、蔡伟平、林新阳、罗雪梅、周泽军、吴树甜、张南宁、陈小珉、杨艳文、韩建强、周定、赵力军、周名嘉、洪东辉、柳建祖、门汉光、屈国伦、郑启皓、周茂、蔡展刚、钟献荣、侯则林、关少梅、黄考颖、黄青、郑晓山、吕向红、罗铁斌、李明、冯志雄、王天、王松帆、陆少芹、庞伟聪、黄勤、丰汉军、郭进军、王维俊、叶充、华锡锋、温武袍、赖海宁、胡世强、陈卫群、郭建昌、李继路、刘谨、王伟、朱宗明、方海龙、郭铭德、黄俊光、黄建华、黄军鹏、黄尚文、颜立华、方良兵、林峻任、朱永东、杨智敏、冯险峰、黄国明、林波、温华威、陈旭、年跟步、胡碧莲、区劲、

their hard work. Hereby I would like to give special thanks to the following partners for their full support: Guo Mingzhuo, Tan Zhuozhi, Zhong Xinquan, Dong Xuekui, Zhao Lu, Chen Shutang, Ma Zhencong, Cai Weiping, Lin Xinyang, Luo Xuemei, Zhou Zejun, Wu Shutian, Zhang Nanning, Chenxiaomin, Yangyanwen, Han Jianqiang, Zhou Ding, Zhao Lijun, Zhou Mingjia, Hong Donghui, Liu Jianzu, Men Hanguang, Qu Guolun, Zheng Qihao, Zhou Mao, Cai Zhangang, Zhong Xianrong, Hou Zelin, Guan Shaomei, Huang Kayoing, Huang Qing, Zheng Xiaoshan, Lv Xianghong, Luotiebin, Li Ming, Feng Zhixiong, Wang Tian, Wang Songfan, Lu Shaoqin, Pang Weicong, Huang Qin, Feng Hanjun, Guo Jinjun, Wang Weijun, Ye Chong, Hua Xifeng, Wen Wupao, Lai Haining, Hu Shiqiang, Chen Weiqun, Guo Jianchang, Li Jilu, Liu Jin, Wang Wei, Zhu Zongming, Fang Hailong, Guo Mingde, Huang Junguang, Huang Jianhua, Huang Junpeng, Huang Shangwen, Yan Lihua, Fang Liangbing, Lin Junren, Zhu Yongdong, Yang Zhimin, Feng Xianfeng, Huang Guoming, Lin Bo, Wen Huawei, Chen Xu, Nian Genbu, Hu Bilian, Ou Jin, Ye Fei, Liang Jundong, Lu Tingkun, Xie Jingwen, Wu Xiling.

I should specially thank the following companies for substantial support:
Guangzhou Design Institute
Guangdong Province Huacheng Architectural Design Co., Ltd.
School of Architecture, South China University of Technology
College of Architecture & Urban Pianning Guangzhou University
Wong Fuksang(France) Urban Planning & Architecture Ltd.
Zhixin High School
Sun Yat-Sen University
Guang Dong Rising Investment Group
Guangdong University of Foreign Studies
Guangzhou Suncity Group
Guangzhou Highsun Enterprises Group Corp
Xinhua Bookstore
Guangzhou Book Center Co., Ltd.
Sunrisinghk Investments Limited
Guangzhou Vocational College of Science and Technology
Acto Real Estate Development Co., Ltd.
Acto (Yangjiang) Wind Power Equipment Manufacturing Co., Ltd.
Guangzhou Suijian Engineering Quality of Safety Testing Center

广州云来斯堡大酒店
Vanburgh Hotel of Guangzhou

三亚市福朋喜来登大酒店
Four Points by Sheraton Sanya

老挝亚欧峰会大酒店
Asia-Europe Summit Hotel, Laos

叶飞、梁俊东、卢挺堃、谢静文、吴喜玲等。

更要感谢各合作公司给予我的大力支持:
广州市设计院
广东省华城建筑设计有限公司
华南理工大学建筑学院
广州大学建筑与城市规划学院
法国黄福生城市规划建筑设计事务所
执信中学
中山大学
广东省广晟投资集团有限公司
广东外语外贸大学
广州太阳城集团
广州海印实业集团有限公司
广州新华书店
广州购书中心有限公司
香港旭日投资集团有限公司
广州科技职业技术学院
广州雅图房地产开发有限公司
广州雅图（阳江）风电设备制造有限公司
广州穗监工程质量安全检测中心
广州云景房地产开发有限公司
广州云星房地产开发集团有限公司
广州金穗实业集团有限公司
广州绿沁度假机构
广州骏发投资有限公司
广州霸王化妆品有限公司
广州勤建置业有限公司
广州津滨·腾越房地产开发有限公司
广州市致友投资有限公司
老挝吉达蓬集团公司
广晟（老挝）投资发展有限公司
柬埔寨加华银行
中工国际工程股份有限公司

Guangzhou Yunjin Real Estate Development Co., Ltd.
Guangzhou Yunxin Real Estate Group Ltd.
Guangzhou Jinsui Industry Group Ltd.
Green Spr Holiday Center
Guangzhou Junfa Investment Co., Ltd.
Guangzhou Bawang Cosmetics Co., Ltd.
Guangzhou Qingjian Properties Co., Ltd.
Guangzhou Jinbin · tengyue Real Estate Development Co., Ltd.
Guangzhou Zhiyou Investment Co., Ltd.
KritTa Phong Group Co., Ltd.
Guangsheng(Laos) Investment & Development (Lao) Co., Ltd.
Canadia Bank of Phnom Penh, Cambodia
China Camc Engineering Co., Ltd.
China National Machinery Industry Corporation
Zhongshan Changzhou Industrial Group Co., Ltd.
People's Government of Huidong County
Kunming Yunhai Environmental Engineering Co., Ltd.
Guangxi Nanning Baiyi Commerciad Co., Ltd.
Nanning Yufeng Real Estate Development Co., Ltd.
Guangxi Shengtian Group
Guangxi Beihai Lvxinxingzhe Real Estate Development Co., Ltd.
Shanxi Donglin Real Estate Development Co., Ltd.
Shanxi Baojie Real Estate Development Co., Ltd.
Jiangxi Xinhaojing Industrial Co., Ltd.
Bureau of Education, Technology and Sports Of Aiyi County
Qingyuan Zhiyuan Real Estate Development Co., Ltd.
Foshan Shibo Real Estate Development Co., Ltd.
Foshan Riches Investment Co., Ltd.
Zhongshan Changzhou Industrial Group Co., Ltd.
Sanjiang Hengyuan Hot Spring Holiday Resort Investment Co., Ltd. Of Lianshan Zhuang and Yao Autonomous County
People's Govemment of Lianshan Zhuang and Yao Autonomous County of Qingyuancity of Guangdong Province
Chongqing Wanda Plaza Properties Co., Ltd.

广州市增城太阳城巧克力社区
Chocolate Community of Sun City at Zengcheng, Guangzhou

广州市花城湾畔
HuaCheng Bay, Guangzhou

广州亚运会亚运村住宅东区规划
Planning of East Residential District for Guangzhou Asian Games, Guangzhou

中国机械工业集团有限公司
中山市长洲实业集团有限公司
昆明云海环境工程有限公司
广西南宁市百益商贸有限公司
南宁裕丰房地产开发有限公司
广西盛天集团
广西北海旅行者房地产开发有限公司
山西东林房地产开发有限公司
山西宝洁房地产开发有限公司
江西新好景实业发展有限公司
江西安义县教育科技体育局
清远市志源房地产开发有限公司
佛山世博房地产开发有限公司
佛山瑞璟投资有限公司
中山市长洲实业集团有限公司
惠东县人民政府
连山县三江恒源温泉度假投资有限公司
连山壮族瑶族自治县（委员会·人民政府）
重庆万达广场置业有限公司
佛山市南海区国土城建和水务局
佛山市南海区国税局
佛山市南海区松岗镇政府
中铁二十五局
中山市中国银行
东兴雄德贸易有限公司
山西湟栋房地产开发有限公司

黄劲
于2014年秋

Bureau of Land and Resources & Municipal and Rural Construction & Water Authority of Nanhai District of Foshan City
Municipal Office of Nanhai District of Foshan City
Songgang Town People's Govemment of Nanhai District of Foshan City
China Railway 25th Bureau Group Co., Ltd.
Bank of China of Zhongshan City
Dongxin Xiongde Commerce & Trade Co., Ltd.
Shanxi Huangdong Real Estate Development Co., Ltd.

Huang Jin
2014 Autumn

城市设计 URBAN DESIGN

广东省肇庆新城市 CBD 核心区规划（鼎湖海印又一城）
The Planning for New Downtown CBD Nodal Region of Zhaoqing of Guangdong (Ding Hu Haiyin Youyicheng)

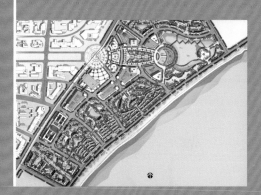

广州大学城组团一的规划设计（中山大学和广东外语外贸大学）
The Planning for the First Group Participation-Guangzhou Higher Education Mega Center (Sun Yat-sen University & Guangdong University of Foreign Studies)

广东省连山小三江温泉度假区的规划设计
The Planning for Xiaosanjiang Hot Spring Resort of Lienshan of Guangdong

广东省连平县南湖蝴蝶谷温泉度假区的规划设计
The Planning for Nanhu Butterfly Valley Hot Springs Resort of Lianping of Guangdong

广州市增城小楼慢城的规划设计
The Planning for Xiaolou Slow City of Zengcheng of Guangzhou

广州市从化小杉村的规划设计
The Planning for Xiaoshan Cun of Conghua of Guangzhou

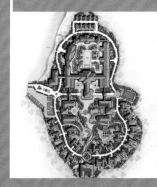

广东省肇庆新城市 CBD 核心区规划（鼎湖海印又一城）

New Downtown CBD Nodal Region of Zhaoqing of Guangdong (Dinghu Haiyin Youyicheng)

项目地点： Venue:	广东省肇庆市鼎湖区 Dinghu District, Zhaoqing, Guangdong Province
设计日期： Design Date:	2011 年 2011
用地面积： Project Area:	898,096 平方米 898,096m²
项目建筑面积： Construction Area:	2,246,330 平方米 2,246,330m²
曾获奖项： Awards:	设计竞赛中标 Bid Winner of Design Competition
合作设计： Partner:	林峻任 Lin Junren

本规划方案力求深层次挖掘肇庆当地文化优势，结合本地段城市配套的功能要求，以及总体规划的空间要求，以抽象的建筑语言，赋予方案独特的文化特点，响应广东省委提出的文化强省的精神。

中轴线以一条跌级水带，连接观砚路及圆形广场，众数座高层商务大楼围绕半月形会展中心布置，抽象再现了包公月牙的传说，喻意"廉政之风皓月可鉴"。

中轴线以超高层超五星级酒店作为标志性建筑，酒店造型有两个寓意。其一作为风帆造型，配合江河环境，有启航之意；其二是造型还让人联想到中国毛笔的笔峰，与砚湖匹配，凸显笔砚文化这一有着更高层次的文化品位背景。

总体规划以包公文化为主题，其半月形会展中心构思源自包公额上月形标志，静夜中月形会展倒影于观砚湖上，"明月为镜，上善若水"，体现了肇庆深厚的文化底蕴。

组团效果图 *Building Cluster of Architectural Renderings*

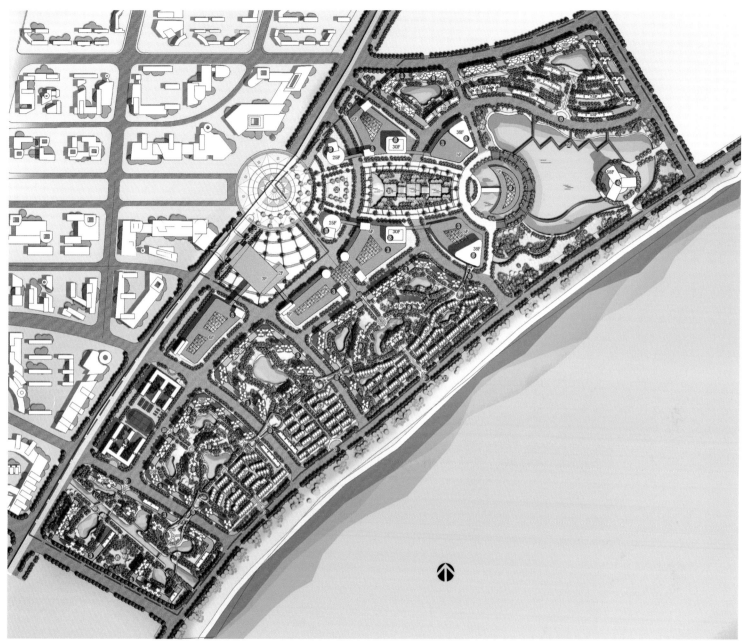

▲规划设计总平面图 *Master Planning + Urban Design*

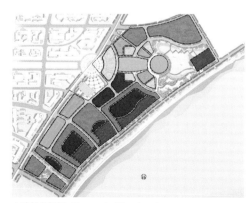

▲功能分析图 *Community Planning plan*

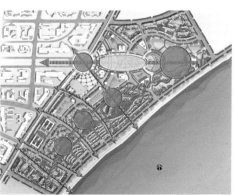

▲景观分析图 *Visual Landscape Analyses Plan*

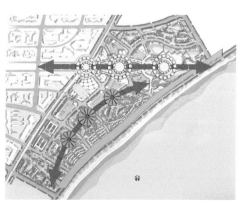

▲规划结构分析图 *Comprehensive and Spatial Planning Plan*

▼鸟瞰效果图 Aerial View

It is the key mission of this project to thoroughly exploit the cultural advantage of Zhaoqing, to combine the functional needs of matching the city properties and space requirement of master planning, hence unique cultural feature is embodied by means of abstract architecture language as a response to the "strengthening of culture" spirit advocated by Guangdong Provincial Committee.

The central axis following a water strip at descending levels links to Guanyan Road and the circular square. Several high-rise business buildings are arranged around the semilunar Exhibition Center, which abstractively matches the legend of Judge Bao's crescent mark and indicates "the bright moon can justify the clean & honest administration".

The high-rise super five-star hotel is the symbolic architecture on the central axis. There are two meanings of the hotel form: firstly, the sail-shape complementing with the river implies "set sail" ; secondly, the building shape also associates with the tip of Chinese brush pen arranged in pairs with the Inkstone (Guanyan) Lake, which outstands the culture of Chinese ancient literature and upgrades the cultural status and taste to a higher level.

The culture of Judge Bao is set for the theme of master planning. The shape of semilunar Exhibition Center deriving from the crescent mark on Judeg Bao's forehead is reflected on the Guanyan Lake in the nighttime. "The bright moon is like the mirror, and the supreme good is like water" , this proverb could best include the profound culture connotation of Zhaoqing.

▲ 鸟瞰效果图 *Aerial View*

▲ 组团内景效果图 *Building Cluster of Architectural Renderings*

▲ 黄劲手绘的酒店建筑设计图 *Huang Jin's Hand Drawing for Hotel Architectural Design*

▲ 酒店建筑效果图 *Hotel Architectural Rendering*

广州大学城组团一的规划设计（中山大学和广东外语外贸大学）

The First Group Participation-Guangzhou Higher Education Mega Center (Sun Yat-sen University & Guangdong University of Foreign Studies)

项目地点： Venue:	广东省广州市小谷围岛 Xiaoguwei Island, Guangzhou, Guangdong Province
设计日期： Design Date:	2003 年 2003
竣工日期： Completion Date:	2004 年 2004
用地面积： Project Area:	180 万平方米 1,800,000m²
项目建筑面积： Construction Area:	80 万平方米 800,000m²
曾获奖项： Awards:	① 国际竞赛第一名 ② 广东省优秀建筑设计一等奖 ③ 广东省优秀规划设计二等奖 ④ 住房和城乡建设部建筑学会创作奖·佳作奖 ⑤ 住房和城乡建设部优秀建筑设计三等奖 ⑥ 中国土木工程詹天佑奖 ① First Prize, International Competition ② First Prize, Excellent Architectural Design of Guangdong Province ③ Second Prize, Excellent Planning & Design of Guangdong Province ④ Honorable Mention Award, Architectural Society of China Architecture Creative Awards, By Ministry of Construction ⑤ Third Prize, Excellent Design of Ministry of Construction ⑥ Tien-Yow Jeme Civil Engineering Prize
合作设计： Partner:	郭明卓、郑启皓、蔡展刚、周茂、钟献荣 Guo Mingzhuo, Zheng Qihao, Cai Zhangang, Zhou Mao, Zhong Xianrong

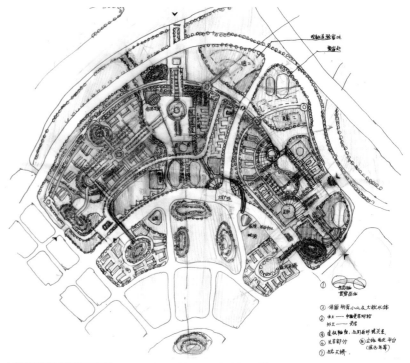

▲黄劲手绘的总规划平面图 *Huang Jin's Hand Drawing for Master Planning*

在新的中山大学和广东外语外贸大学的规划设计中，我们以岭南生态校园环境和原中山大学校园文化为创作主题，以自然环境的保护和利用为构思出发点，将主体构思贯穿于整体布局、单体及细部的系统设计中，创造出具有岭南特色的生态校园环境。

设计方案尽一切可能不破坏用地内的绿化、山体，基本保留原有山体水系，并对这一自然现状加以再利用，形成校园中现有的主要景观。总平面布局借用"曲水流觞"意境来组织生态轴。整个校园对现有水系进行整合，以生态轴串起几十个园林空间。生态轴以其自由优雅的曲线空间，从南到北、由东向西贯穿整个校园，水的浪漫气息穿行于校园建筑之间。

中山大学主广场及入口位于大学城中心的正北向，此处江面最为开阔。由图书馆、公共教学楼、行政会议中心围合的建筑群体是中大校园标志性建筑群。以孙中山像为中心的几何图案化的广场绿化，与开阔的珠江美景，给人以气势宏伟的感受，体现中山大学治学严谨、勇于开创的学术氛围。

传媒学院、基础实验楼、工学院等教学楼围绕生态轴临江布置。教学区尽末端的工学院，以标志塔再现原中大钟塔的风貌，是教学区及生活区轴线交会处的视线焦点。建筑群具有错落起伏的轮廓线。

外墙色彩以原校本部南区建筑的棕红色调为主色，山墙组合以白色景墙为背景，突出红色的墙面入口造型，形成控制整体风格的母题，结合轻质钢结构及通透玻璃等现代材料，造型具有雕塑美感及时代气息，体现中山大学校园浓厚的文化底蕴。

▲ 规划设计总平面图 *Master Planning + Urban Design*

Lingnan ecological campus environment and original campus culture of Sun Yat-sen University are the main creative themes of the planning and design of new Sun Yat-sen University and Guangdong University of Foreign Studies. Inspired by protection and utilization of natural environment, the key concept emerges throughout the systemic design of the entire layout, individual units and details, resulting in the ecological campus environment full of characteristic lingnan style.

The current major landscape is engendered without destroying the plants and hills within the campus, maintaining the existing landscape and waterscape and reutilizing the natural environment. The artistic conception of floating "Wine Cup along the Winding Water" is employed to arrange the ecological axis of the master layout. The existing water system on the campus is integrated to link together tens of gardens with ecological axis, which goes through the entire campus from south to north and from east to west in a free-style and graceful curve. All buildings on the campus are immersed in the romantic mood of waterscape.

The main square and entrance of Sun Yat-sen University are located at the north of the center of Guangzhou Higher Education Mega Center, with an open view of the Pearl River. The building complex enclosed by Library, Public Classroom Building, Administration and Conference Center turns out to be the symbolic architectures of the University. The geometric greenery on the square centered by the statue of Mr. Sun Yat-sen work together with the vast scenery of the Pearl River, giving impressive and imposing aura, while bringing forth the academic atmosphere of rigorous scholarship and courageous innovation of Sun Yat-sen University.

School of Media and Communications, Fundamental Laboratory Building and School of Polytechnic are deployed around the ecological axis by the River. The School of Polytechnic at the end of the Classroom Area forms the visual focus at the axis joint of Classroom Area and Living Quarters. The building complex has undulate and irregular silhouette.

Reddish brown of the buildings on original south campus becomes the key tone of the exterior, white landscape wall sets the background of the gable wall combination and accentuates the red color and entrance shape of the wall, forming the master theme deciding the holistic style. Joined with modern materials such as light-weight steel structure and transparent glass etc., The form is full of sculptural beauty and contemporary touch, which reflects the connotation of profound culture of Sun Yat-sen University.

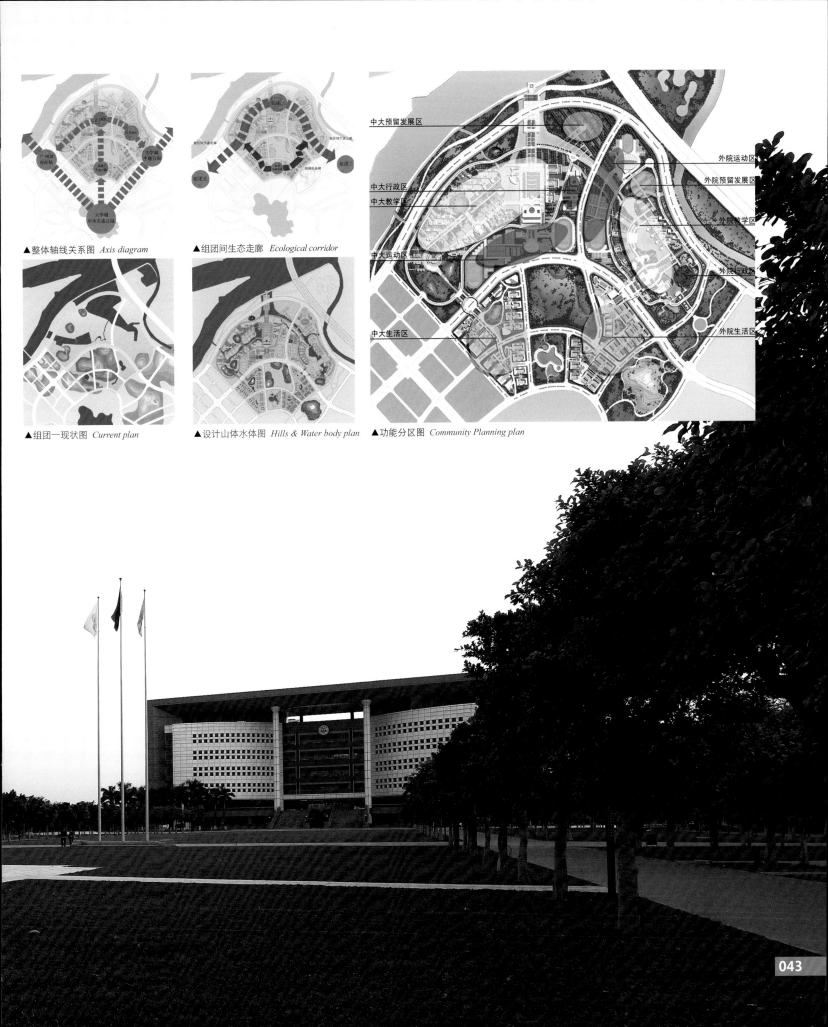

▲ 鸟瞰效果图 *Aerial View*

047

广东省连山小三江温泉度假区
Xiaosanjiang Hot Spring Resort of Lianshan of Guangdong

项目地点： Venue:	广东省连山壮族瑶族自治县小三江镇 Lianshan Zhuang and Yao Autonomous County, Guangdong Province
设计日期： Design Date:	2013 年 2013
用地面积： Project Area:	1,333,333 平方米（2000 亩） 1,333,333m² (2 000a)
合作设计： Partner:	李滔、林波 Li Tao, Lin Bo

项目位于连山南麓小三江镇，用地四面环山，有一条溪水自北向南流过，设计上充分利用优美的地形环境，利用溪水构成生态水轴，作为规划结构的主要骨架，所有建筑组成均围绕水轴构成，形成生态有序的整体。

规划包括温泉酒店、步行商业街、度假别墅、森林公园、生态水系等休闲项目。

建筑风格采用当地特有的壮瑶建筑特色，以壮瑶文化为主题构思，酒店前的湖面和舞台特为泼水节而设计。

Lied in Xiao Sanjiang Town in the south foothill of Lianshan, Xiao Sanjiang Resort & Spa is surrounded by mountains with a stream running through from north to south. The designer takes full advantage of the beautiful landform, using the stream to form the ecological water axis as the main skeleton of planning structure, thus all buildings are built around the water as a whole in perfect ecology and good order.

The design includes recreational projects such as spa hotel, pedestrian commercial street, holiday villas, forest park and ecological water system etc.

The design features indigenous architecture style of Zhuang and Yao Nationality and evolves around the theme of their culture. The lake and stage in front of the hotel is specially made for the Water-splashing Festival.

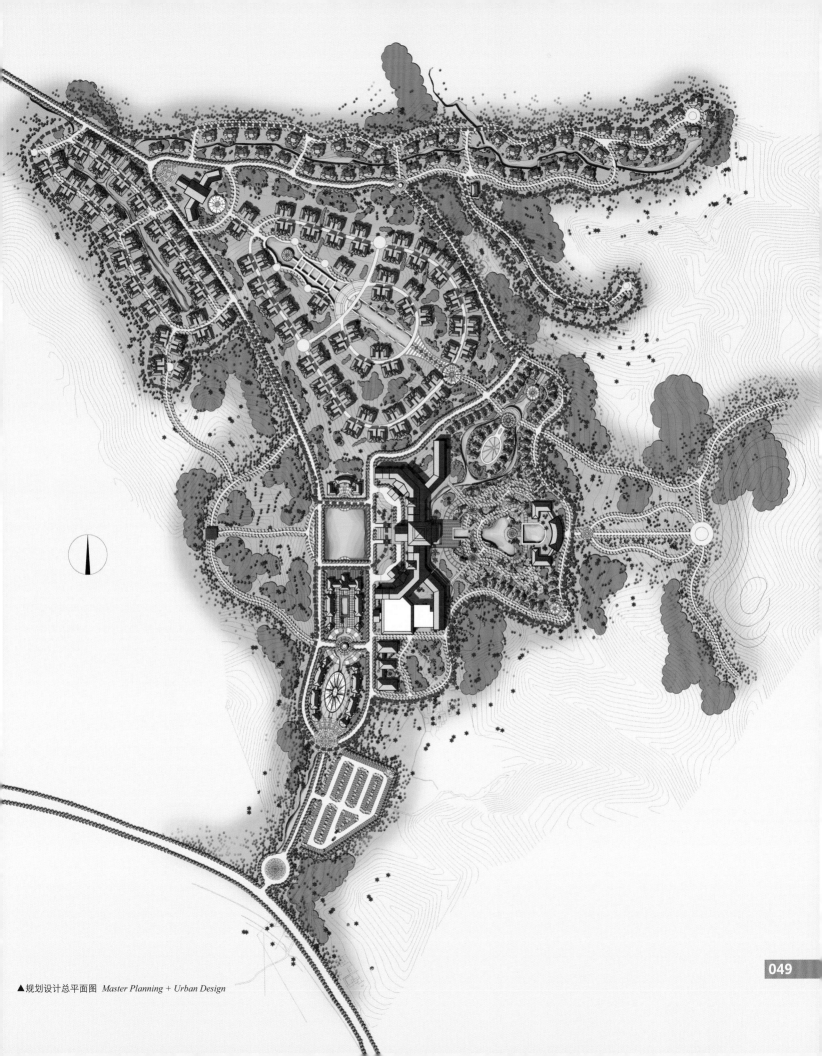

▲ 规划设计总平面图 *Master Planning + Urban Design*

广东省连平县南湖蝴蝶谷温泉度假区
Nanhu Butterfly Valley Hot Springs Resort of Lianping of Guangdong

项目地点： Venue:	广东省河源市连平县南湖镇 Nanhu Town, Lianping County, Heyuan City, Guangdong Province
设计日期： Design Date:	2013 年 2013
用地面积： Project Area:	2,000,000 平方米（3 000 亩） 2,000,000m²(3 000a)
合作设计： Partner:	林波 Lin Bo

连平县南湖蝴蝶谷温泉度假区项目占地 200 万平方米，位于和平县南湖村附近的高山上，是国内罕有的高山温泉度假区。

在项目设计中，我们充分利用当地优美的自然生态环境，筑坝储水，形成高山湖泊，使之成为景区的中心景观。整体布局以一条生态旅游文化轴为主体结构，轴线上贯通入口、各个景区、湖泊，最终以阳明书院为轴线空间序列的高潮，突出当地独有的"王阳明"文化特点。

建筑围绕两环景观带布置，内环为环湖景观带，外环为山林景观带，建筑靠山面水，尽览大自然美景。

Being one of the rare high-hill resort and spas in China, Nanhu Butterfly Valley Resort and Spa is perched on the mountain near the Nanhu Village of Lianping County with a total area of 3,000a.

In this project, designer has fully utilized the beautiful natural environment, and built the dam to store water, the mountain and lake as a result become the central view of all scenic spots. The overall layout is centralized on the ecological tourism cultural axis, on which all entrances, scenic spots and lakes are connected and reached to climax in axis space sequence with Yangming Academy, that also outstands the particular cultural feature of "Wang Yangming".

The buildings are set around two rings of landscape: the inner ring is the landscape belt around the lake, and the outer ring is the landscape belt of the forest. This design enables the beautiful scenery of nature to be enjoyed from the buildings fronting water and leaning against the hill.

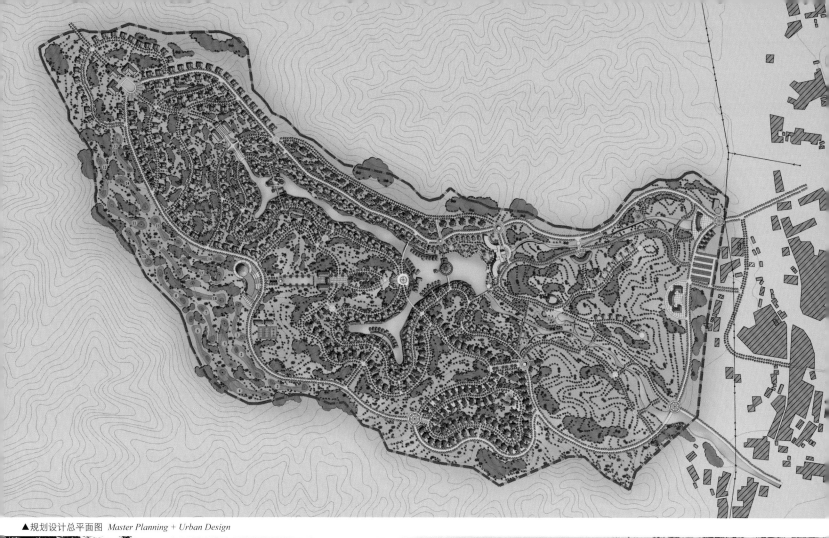

▲规划设计总平面图 *Master Planning + Urban Design*

广州市增城小楼慢城
Xiaolou Slow City of Zengcheng of Guangzhou

项目地点： Venue:	广东省广州市增城小楼镇 Xiaolou, Zengcheng, Guangzhou City, Guangdong Province
设计日期： Design Date:	2013 年 2013
用地面积： Project Area:	200 万平方米 2,000,000m²
合作设计： Partner:	林波 Lin Bo

▲ 鸟瞰效果图 *3D Rendering*

本方案以慢城的国际标准打造广东首个"慢城"，慢城的核心是倡导纯粹的生活，保护当地特色，对伴随全球化而来的同质化和标准化说不，这种"由快到慢"的精英生活模式将很快风靡全球。小楼慢城项目强调人与自然的和谐发展，强调生活的品质，强调在悠闲的生活节奏中回归生活的本质，体会生命的意义。规划上力求深层次挖掘增城当地的生态和文化优势，结合本地段城市配套的功能要求，以及总体规划的空间要求，以抽象的建筑语言，赋予方案独特的文化特点，应增城市政府对本地段旅游观光、生态农业的要求，打造4A景区，打造生态"文明之城"、"小康之城"、"最具幸福感之城"。

本区域以一条水系贯通各区，水在古代文化中代表了安宁、丰收和幸福，充分利用原地形中这丰富的水资源，是打造具有岭南特点的水城，打造宜居、宜游、宜养的幸福之城之关键。

规划结构呈一心二环三轴形式，重点打造的旅游生态轴，将各个功能地块连为一体，生态轴自A区高山瀑布而下，沿湖岸至B区中心区域，再延伸至C、D区的生态水域，生态轴上是一系列旅游项目和旅游景区，漫步其中，尽享湖光山水之美景。

The project is dedicated to create the first "slow city" in Guangdong at international standard. "Slow city" is cored on advocating pure life, protecting local speciality and saying "no" to homogenization and standardization coming along with globalization, with its "from fast to slow" elite lifestyle it will spread all over the world soon. "Small Building & Slow City" attaches much importance to the harmonious development of human and nature, and the quality of life as well, which encourages people to return to the real nature and experience the meaning of life in leisurable pace. It works best to deeply exploit the advantages of local ecology and culture of Zengcheng, and combines the functional requirement of local urban infrastructure and space requirement of master planning to endow unique cultural feature to this project. As per the request of Zengcheng Municipal Government on local tourism and ecological agriculture, to create 4A Grade Tourist Attraction of "civilized city", "comfortably-off city" and "happiest city".

The site is penetrated by a water system connecting to all areas. Water represents peace, harvest and happiness in ancient culture. It is key to build the water city of Lingnan features and happy city suitable for residing, sightseeing and nurturing in full utilization of rich water resource of the original landform.

The structure is planned in the form of "one core, two rings and three axes". The tourism ecological axis is specially designed to link each functional land together, which falls down with the waterfall from high hill in A Zone, reaches to the central area of B Zone along the lakeside, and extends to the ecological water areas of C & D Zones. There are a series of scenic spots and sightseeing points on the ecological axis, people could enjoy the beauty of lake, mountain and water by wandering amongst the picturesque scenery.

▲ 规划设计总平面图 *Master Planning + Urban Design*

津滨·腾越大厦
Jinbin·Tengyue Mansion

勤建大厦
Qinjian Building

佛山市世博商业中心
World Expo Business Center of Foshan

佛山市南海区国土局地籍资料库
Cadastral Database of Land and Resources Bureau of Nanhai District of Foshan

公共建筑 PUBLIC BUILDINGS

广州购书中心及维多利广场
Guangzhou Book Center & Victory Plaza

广州大学城购书中心
Book Center of Guangzhou Higher Education Mega Center

新华大厦
Xinhua Building

佛山市松岗广场
Songgang Square of Foshan

佛山市南海区国家税务局综合业务用房
Integrated Service Office of State Administration of Taxation of Nanhai District

南宁市云星办公楼
Yunxing Office Building of Nanning

惠州市惠东县博物馆
Huidong County Museum of Huizhou

中山市中国银行大厦
Bank of China Tower of Zhongshan

南方铁道大厦
South Railway Building

广东金融高新技术服务区 C 区商业城
Commercial City of Area C at the Guangdong High Tech Service Zone for Financial Institutions

十三行国际时装城（广州市荔湾区西关旧城改造）
Thirteen Hong International Fashion City (Xiguan Urban Redevelopment Project of Liwan of Guangzhou)

老挝万象湄公河畔商业步行街西区
The West of Mekong River Commercial Pedestrian Street at the Vientiane, Laos

重庆巴南万达广场
Chongqing Banan Wanda Plaza

柬埔寨金边加华银行
Canadia Bank of Phnom Penh of Cambodia

太原艺术博物馆
Taiyuan Museum of Art

英格（阳江）电气有限公司生产基地
Production Sites of Yangjiang ENGGA Generators Co., Ltd.

广州市致友鲜肉冻品交易市场
Guangzhou Zhiyou Fresh frozen food trading market

津滨·腾越大厦
Jinbin · Tengyue Mansion

项目地点： Venue:	广州市珠江新城 Zhujiang New Town, Guangzhou
设计日期： Design Date:	2006 年 2006
竣工日期： Completion Date:	2008 年 2008
项目建筑面积： Construction Area:	83,200 平方米 83,200m²
用地面积： Project Area:	12,000 平方米 12,000m²
曾获奖项： Awards:	① 住房和城乡建设部优秀建筑设计二等奖 ② 广东省优秀建筑设计二等奖 ① Second Prize, Excellent Design of Ministry of Construction ② Second Prize, Excellent Architectural Design of Guangdong Province
合作设计： Partner:	黄考颖、郑蕊、黄天 Huang Kaoying, Zheng Rui, Huang Tian

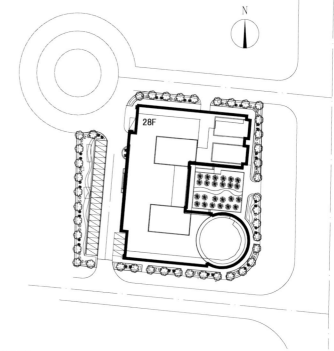

▲总平面图 *Master Plan*

▲黄劲手绘效果图 *Hand Drawing, Huang Jin*

　　津滨·腾越大厦本大厦位于广州中轴线珠江新城 CBD 地段，是一座具有新岭南风格的智慧型、生态型、节能型之现代化办公楼。

　　大楼平面顺应主导风向，围绕内庭园呈凹字形布置，在风流方向上分层设置多个导风口，使办公用房有很好的自然通风采光。

　　大楼采用立体绿化设计，首层的入口庭园、每五层设置空中庭园以及众多绿化平台，创造了很好的微气候环境。外形以方、圆形体组合，取"天圆地方"之寓意，外形刚柔并济，挺拔俊朗。立面采用竖向分段式构图，细部处理精致。

　　大厦在多项建筑技术上探求创新，立面采用"仿玻璃幕墙效果"设计（如使用建筑外墙氟碳面漆），利用窗体结构将建筑幕墙的塑造与窗的造型融为一体，以普通窗的造价设计出铝合金玻璃幕墙的立面效果。在广州首次采用建筑幕墙内开启窗设计，解决了幕墙内渗水的技术问题，避免窗户外开启对建筑外立面的影响。

　　以"二十四小时办公楼"是该大厦的一个重要理念，结合创业型写字楼全天候办公的特点，大厦采用全中央空调、分层小中央空调、分体空调共用形式，适应各种条件下客房租户的办公要求，同时合理控制空调能耗。

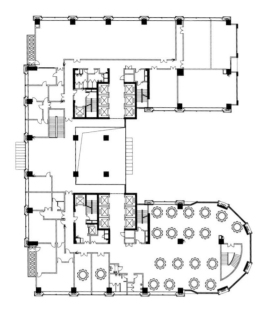

▲ 二层平面图 Level Two Plan

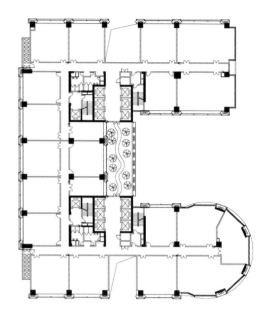

▲ 标准层 -2 平面图 Standard Floor Plan -2

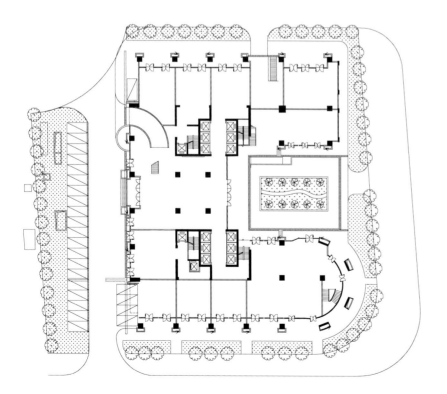

▲ 首层平面图 Ground Level Plan

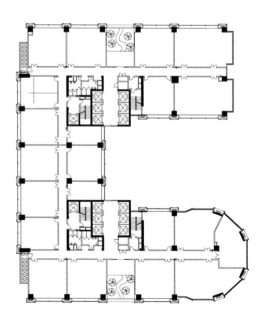

▲ 标准层 -1 平面图 Standard Floor Plan -1

Jinbin · Tengyue Mansion sits on the CBD of Zhujiang New Town—the central axis of Guangzhou. It is an intelligent, ecological, energy-saving and modern office building of neo-Lingnan style.

The building is planned in the shape of Chinese character "凹" around the inner courtyard in accord with wind direction. In the direction of wind a couple of air ducts are installed in layers to enable excellent ventilation and light for the office.

As far as three-dimensional green design is introduced, the entrance courtyard, the hanging gardens on every five stories and numerous green platforms create favorable microclimate environment. "The dome-like heaven embraces the vast earth" is indicated for the combination of square and circle in the form, which is upright and beautiful in perfect match of hardness with softness. The elevation is structured in vertical-sectional way with delicate manipulation in details.

Innovation is exploited on several architectural technologies for the Building. The "glass curtain wall-imitating" design (e.g. use of fluorocarbon coatings for the exterior) of elevation merges the windows into the curtain wall with window structure, the visual effect of curtain wall made of aluminum alloy is brought out at the budget of ordinary window. Inner-open window design is first adopted in China for curtain wall, which serves to solve the technical problem of dampness penetration and avoid the influence on elevation look by opening the window outwards.

"Around-the-clock office building" being the important philosophy, in combination with the feature of all-day innovative entrepreneurial office building, Jinbin · Tengyue cleverly adheres to the mixed usage of central air-conditioning, independent central air-conditioning on different stories, and separated air conditioners. This practice could meet all requirements of different tenants in various conditions, meanwhile reasonably control the energy consumption of air conditioning.

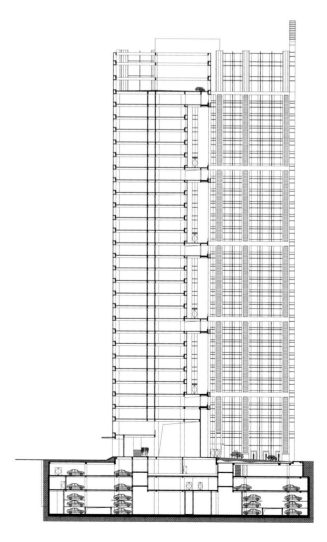

▲ 剖立面图 *Section*

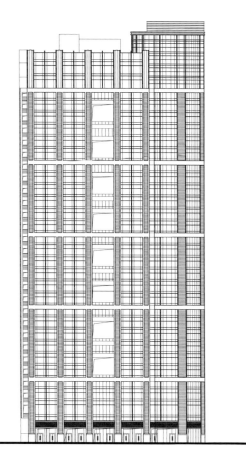
▲南立面图 *South Elevation*

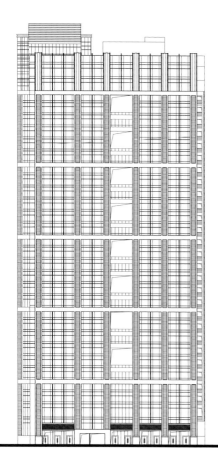
▲北立面图 *North Elevation*

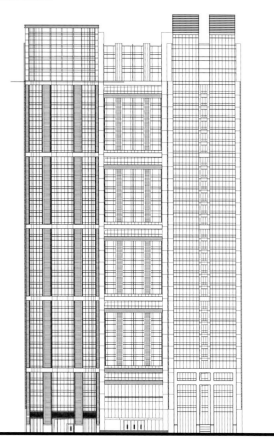
▲东立面图 *East Elevation*

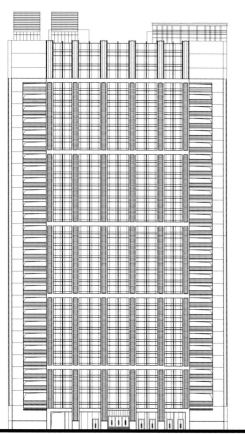
▲西立面图 *West Elevation*

勤建大厦
Qinjian Building

项目地点： Venue:	广州市珠江新城 Zhujiang New Town, Guangzhou
设计日期： Design Date:	2001 年 2001
竣工日期： Completion Date:	2003 年 2003
项目建筑面积： Construction Area:	23,800 平方米 23,800m²
用地面积： Project Area:	8,200 平方米 8,200m²
曾获奖项： Awards:	广东省优秀建筑设计三等奖 Third Prize, Excellent Architectural Design of Guangdong Province
合作设计： Partner:	江涛 Jiang Tao

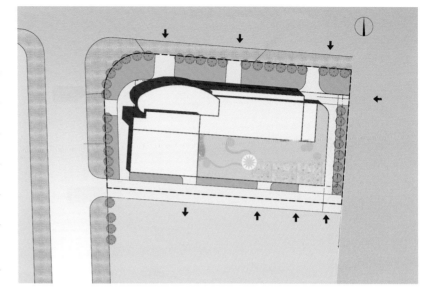

▲总平面图 *Master Plan*

本项目是具有岭南特色的生态型甲级商务写字楼，其平面在路口转弯角作弧形变化，巧妙地将平面切割成方、圆两个体块，中央以多个空中花园相连，每个办公单元均可观赏园林花园景致，形成大楼设计上的亮点，大楼以其俊秀的造型及空中花园设计，在新城市中心的时尚办公楼设计中独树一帜。

Qinjian Building is an ecological Grade-A office building of Lingnan style. Curvy change is designed for the corner at the crossing skillfully cutting the planed into two parts in square and circle. With connection to various hanging gardens in the center, each office can share the beautiful garden view, which is the quintessence of the design. The handsome shape and hanging garden design enable the Building to stand out from the contemporary office designs in the new city center.

▲黄劲手绘效果图 *Hand Drawing, Huang Jin*

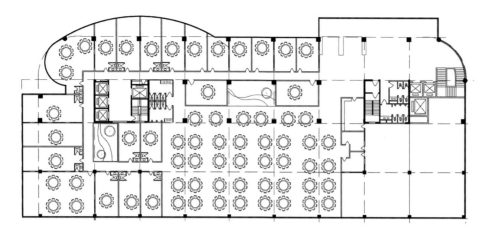

▲三、四层平面图 *Level Three & Four Plan*

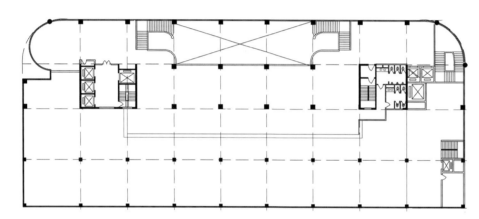

▲二层平面图 *Level Two Plan*

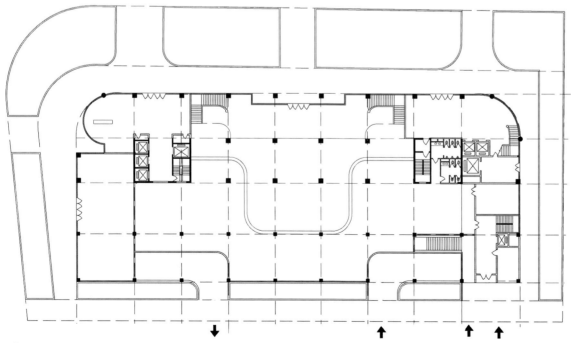

▲首层平面图 *Ground Level Plan*

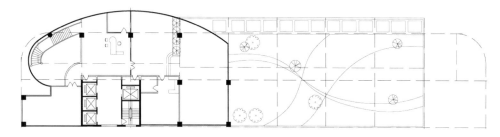

▲二十五层平面图 *Level Twenty Five Plan*

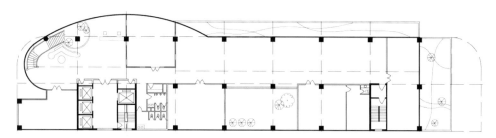

▲二十四层平面图 *Level Twenty Four Plan*

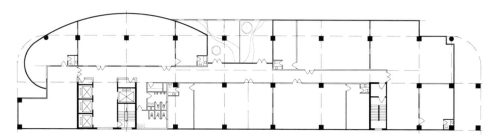

▲标准层平面图 *Standard Floor Plan*

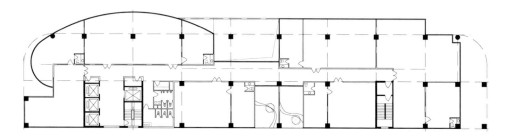

▲标准层-1平面图 *Standard Floor Plan-1*

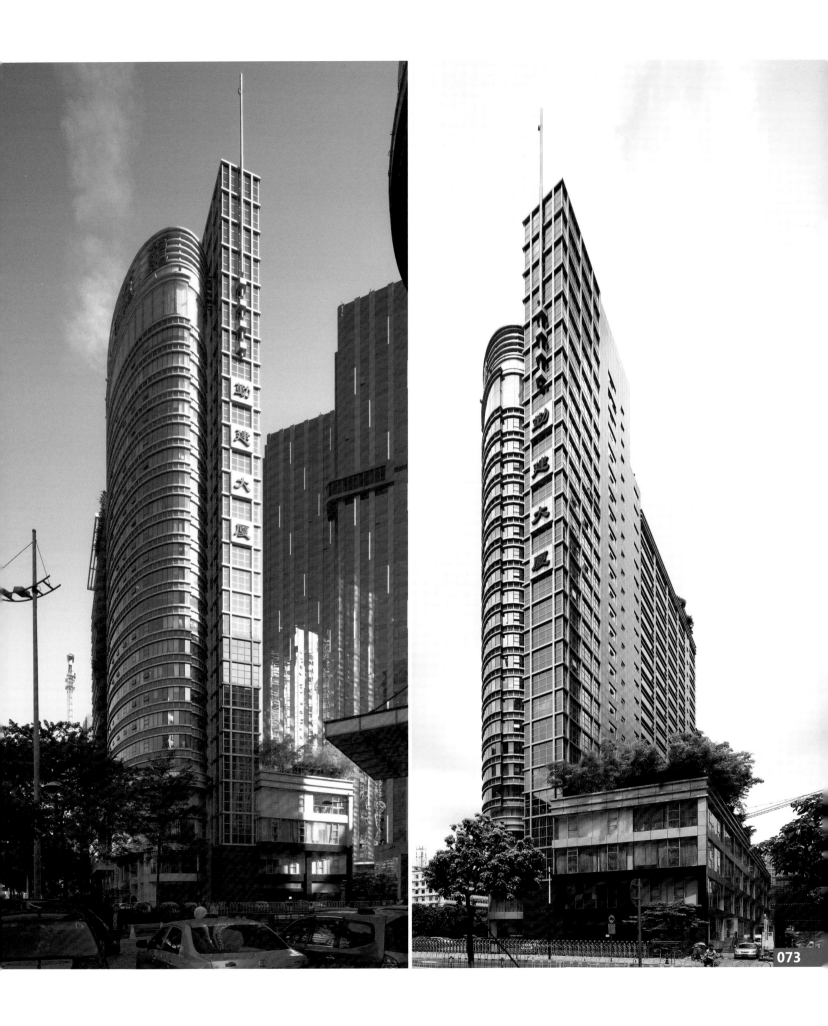

广州购书中心及维多利广场
Guangzhou Book Center & Victory Plaza

项目地点： Venue:	广州市天河区 Tianhe District, Guangzhou
设计日期： Design Date:	1992 年 1992
竣工日期： Completion Date:	1994 年 1994
建筑面积： Project Area:	250,000 平方米 250,000m²
曾获奖项： Awards:	① 广东省优秀建筑设计一等奖 ② 住房和城乡建设部优秀建筑设计二等奖 ③ 国家优秀设计铜奖 ① First Prize, Excellent Architectural Design of Guangdong Province ② Second Prize, Excellent Design of Ministry of Construction ③ Bronze Award, State Excellent Design
合作设计： Partner:	郭明卓、陈树棠 Guo Mingzhuo, Chen Shutang

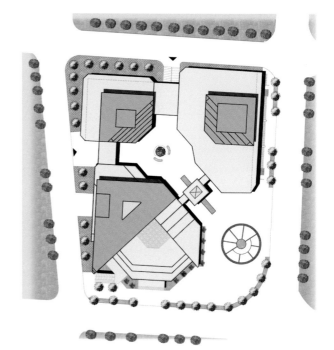

▲ 总平面图 *Master Plan*

广州购书中心将商业性及艺术性融合为一体，既要创造良好的购书氛围及巨大的经济效益，又要充分体现其文化建筑的艺术特性，广州购书中心在这些方面作了探索。广州购书中心位于主干道天河路，东邻天河体育中心。广州购书中心项目用地约 6,000m²，建筑面积为 23,000m²，地下层为设备车库，一～五层为营销、批发的场地，五～九层为办公及车库。

购书中心在设计上力求综合考虑平面布局、结构、外形等关系，以简洁的结构平面布局，充分满足使用要求，形体自然，丰富而具个性。

1. 总体规划

购书中心与相邻两座高层呈品字形组合，在路口交会处以一条 45°度灵活对称的轴线控制全局，与东邻的天河体育中心对称性布局互相响应。

2. 中庭

裙房围绕中庭布局，中庭以一面三层高的浮雕作为入口处对景，中庭层层跌级，暴露结构柱，多层空间，自然穿插，丰富而有序，自动扶梯中庭边上布置，可观赏各层景色。中庭视野开阔，形成空间重点，创造良好购书气氛及艺术氛围。

3. 结构与外形

结构采用 8m×8m 方格平面，利用 45°度切角自然变化出丰富的立面，但其使用空间仍为以方方为主。其立面变化没有影响平面空间的使用。

4. 外形采用象征主义手法

塔楼自然张开，以利于塔楼南北向通风采光，外形犹如一本张开翻开的图书。裙房层层跌级，以减少对天河路的压迫感，同时寓意"循序渐进"的学习方法，其穹形大门及顶部圆窗组合暗示"知识宝库的大门及钥匙"。

维多利广场是大型的城市综合体，是与广州购书中心作为一组整体建筑群而进行规划设计的，规划结构源自用地东南角的城市广场及其 45°轴线，总平面顺应轴线形成的 Y 形商业步行街。三座建筑拟呈品字形布局，维多利广场划分为两座超高层甲级写字楼，其方形平面在顶部作跌级变化，结合升起的交通核心，外形新颖独特，外形设计以"钻石"立意，在外形上做多个方向切割，在不同光影下展现其晶莹亮泽的钻石造型，是新城市中轴线上的一组标志性建筑。

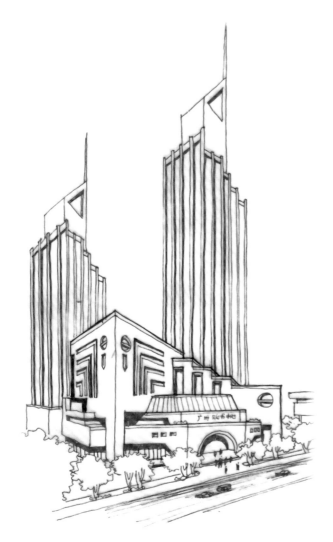

▲ 黄劲手绘效果图 *Hand Drawing, Huang Jin*

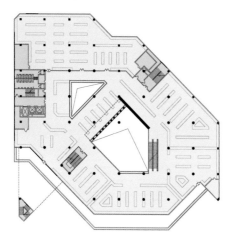

▲ 三层 Level Three Plan

▲ 二层 Level Two Plan

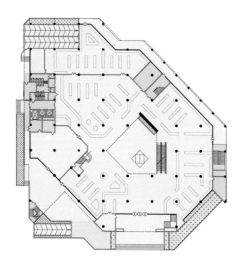

▲ 一层 Level One Plan

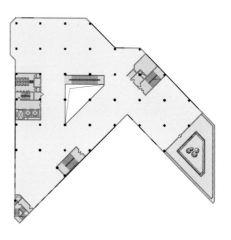

▲ 六层 Level Six Plan

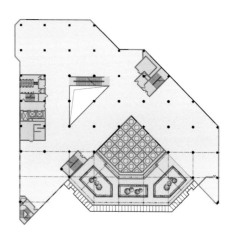

▲ 五层 Level Five Plan

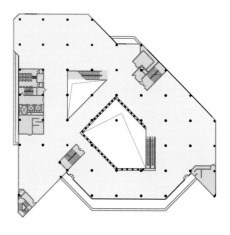

▲ 四层 Level Four Plan

Integrated commerce with art, Guangzhou Book Center creates a nice environment for bookstores, as well as enormous economic value, which is fully expressed within its artistic and cultural architecture. Guangzhou Book Center has explored all these aspects. It sits on the arterial street of Tianhe Road and besides Tianhe Sports Center to the east. With the Guangzhou Book Center total Project Area of 6,000m², and construction area of 23,000m². Guangzhou Book Center accommodates the equipment car park in the basement, retail and wholesale space on the 1st ~ 5th floor, and office and car park on the 5th to 9th floor.

The design has comprehensively considered the relationship amongst the layout, structure and shape etc., making it simple and clean enough to fulfill the functional requirement and create a natural, rich and characteristic form.

1. Master Plan

Guangzhou Book Center and two adjacent high-rises compose the shape of Chinese character "品", meeting at the intersection to engender a flexible and symmetrical axis at an angle of 45 degrees, in response to the symmetrical layout of Tianhe Sports Center to the east.

2. Atrium

The podium encircles the atrium which sets the opposite scenery at the entrance by a bas-relief wall of three-story high. The atrium is in split levels with exposed structure columns naturally interwoven in layers of spaces, giving a rich and organized feeling. The escalator at one side of the atrium enables readers to freely enjoy the view of each floor. The open view of the atrium as the focal point brings out favorable reading and artistic atmosphere to the space.

3. Structure and Form

The structure of 8m x 8m square grid presents a rich elevation with the natural change caused by 45 degree angle. The plane still takes square shape for functional space in principle without any interference from the elevation change.

4. Shape of Symbolism

The Tower Building naturally unfolds like an open book to admit natural light and ventilation in south-north direction. The podium descends in steps to release the pressure on Tianhe Road implying the study method of "improvement in proper sequence". The arc-shape gate and circular window atop also give a hint of "the gate and key to the thesauru".

Victory Plaza is a large metropolitan complex planned and designed with Guangzhou Book Center as a complete group of architecture. The structure uses the city plaza on the southeast corner and its 45 degree axis, the pedestrian street in the shape of "Y" is formed in compliance with the axis, and the three buildings is set in the layout of Chinese character "品". Victory Plaza is composed of two high-rise Grade-A Office Buildings, with the square plane descends in steps at the top, in combination with the rising traffic core, resulting in a unique and novel look. The architecture shape is enlightened by "diamond" with incisions from multi-directions, glittering the crystal-like shine of the diamond in different lights and shadows, which gives itself prominence as the symbolic architecture of this new central axis of the city.

广州大学城购书中心
Book Center of Guangzhou Higher Education Mega Center

项目地点： Venue:	广州大学城 Guangzhou Higher Education Mega Center
设计日期： Design Date:	2002 年 2002
竣工日期： Completion Date:	2004 年 2004
项目建筑面积： Construction Area:	20,000 平方米 20,000m²
用地面积： Project Area:	12,000 平方米 12,000m²
曾获奖项： Awards:	设计竞赛中标 Bid Winner of Design Competition
合作设计： Partner:	郑晓山 Zheng Xiaoshan

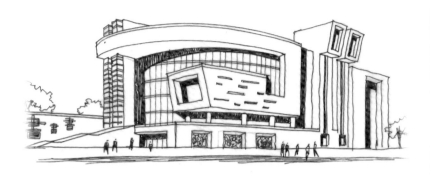

▲ 黄劲手绘效果图　*Hand Drawing, Huang Jin*

本项目位于大学城中轴线南端的珠江河畔，主要功能为图书营销和发行，并作为大学城公众活动场所。

项目以"知识方舟"作为主体构思，具有雕塑感的造型，体现了购书中心的文化特征。购书中心柱网规整，内部空间围绕中庭布置，中庭以中国古典文字作为景墙装饰，其升高平台平时可举行多种文化活动。

购书中心外部空间犹如一个抽象的大门，门框内的书卷形的弧面极具个性，购书中心主入口门厅设于二层，层层向上的台阶诠译了"书山有路勤为径，学海无涯苦作舟"的名句。购书中心塔式造型及入口的方形口部造型艺术地抽取了中国古典印章造型，其雕塑造型具有文化意味。

The project situates by the Pearl River at the south end of the central axis of Guangzhou Higher Education Mega Center. The Book Center serves for marketing and distribution of books as well as the public activity center.

"Ark of Knowledge" is taken as the main concept hence the sculptural shape nicely represents the cultural feature of the Book Center—the column network is regular; the inner space is arranged around the atrium which is decorated with traditional Chinese characters as landscape wall; the raised platform can accommodate various cultural events.

The exterior of the Book Center is like an abstract door with characteristic scroll-shape cambered surface in the doorframe. The main entrance hall is on the second floor. The ascendant steps signify the famous sentence of poetry—there is no royal road to learning. The sculptural and cultural tower-form of the Book Center and the square mouth-shape of the entrance are artistically extracted from the conformation of Chinese classical seal.

▲四层平面图 Level Four Plan

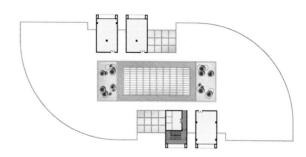

▲屋面平面图 Roof Level Plan

▲三层平面图 Level Three Plan

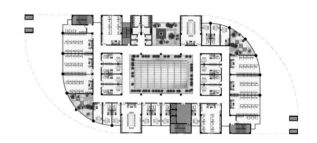

▲六层平面图 Level Six Plan

▲二层平面图 Level Two Plan

▲五层平面图 Level Five Plan

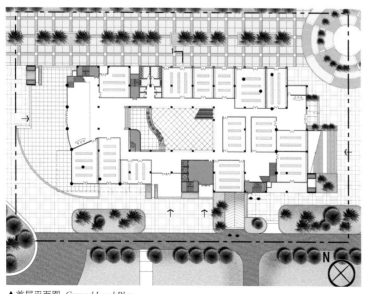

▲首层平面图 Ground Level Plan

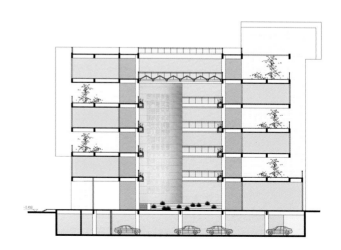

▲1-1 剖面图 1-1 Section

新华大厦
Xinhua Building

项目地点： Venue:	广州市小北路 Xiaobei Road, Guangzhou
设计日期： Design Date:	2006 年 2006
竣工日期： Completion Date:	2010 年 2010
项目建筑面积： Construction Area:	15,893 平方米 15,893m²
用地面积： Project Area:	2,710 平方米 2,710m²
曾获奖项： Awards:	广东省优秀建筑设计二等奖 Second Prize, Excellent Architectural Design of Guangdong Province
合作设计： Partner :	郑晓山、关小梅 Zheng Xiaoshan, Guan Xiaomei

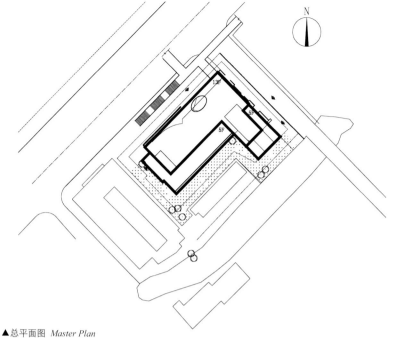

▲总平面图 *Master Plan*

新华大厦是一座商业性建筑，而其商品却是具有文化艺术性的图书及音像制品。因此，在设计上考虑购书气氛及经济效益的同时，还要力求创造良好的艺术气氛，融商业性与文化艺术性为一体。

在设计中，我们引入生态建筑的理念，创造良好的自然通风采光，节能省耗，以立体绿化创造优美的购书、办公环境。

建筑外形有强烈的虚实变化，以通透玻璃向路人展现其商业形象，以雕塑感极强的实面体现其艺术性。外形设计以"书库"为主线，具有独特的形象。整体外形呈一座书库，喻意书的容器，容器正面西侧自然呈弧面张开，犹如打开的书页，正面层层退级及侧面竖向板块造型，也让人联想到横竖放置的书本。顶部屋面花园由一折板遮盖，产生独特的空间效果，中设圆孔可引入灵光，造型活泼、有变化。主体入口趋向于东北角，以避开天桥对大楼的影响。

Xinhua Building (Mansion) is a commercial building, with full display of books and audio products regarding culture and art. Hence the design is required to consider shopping atmosphere and economic benefit meanwhile create favorable artistic mood and faultlessly integrate business, art and culture.

The design introduces the concept of ecological architecture, with fine and natural ventilation and lighting, saved and effective energy and power, creating excellent shopping and office environment by vertical planting.

There are strong virtual and solid changes in the architecture form. The transparent glass displays the business image to the passerby, and the highly sculptural solid surface represents the sense of art. The unique image of "library" is the spine of the shape, which implies the container of books. The west side of elevation stretches in natural curvy way as if an open book; the recessed steps on the facade and the vertical plate-form on the side remind of the books placed either horizontally and vertically. The roof-top garden covered by folded plate creates special space effect, the circular hole in the middle introduces aerial light, all this work together to give a vivid and changeful form. The main entrance tends to the northeast corner could avoid influence to the building from the sky bridge.

▲ 黄劲手绘效果图 *Hand Drawing, Huang Jin*

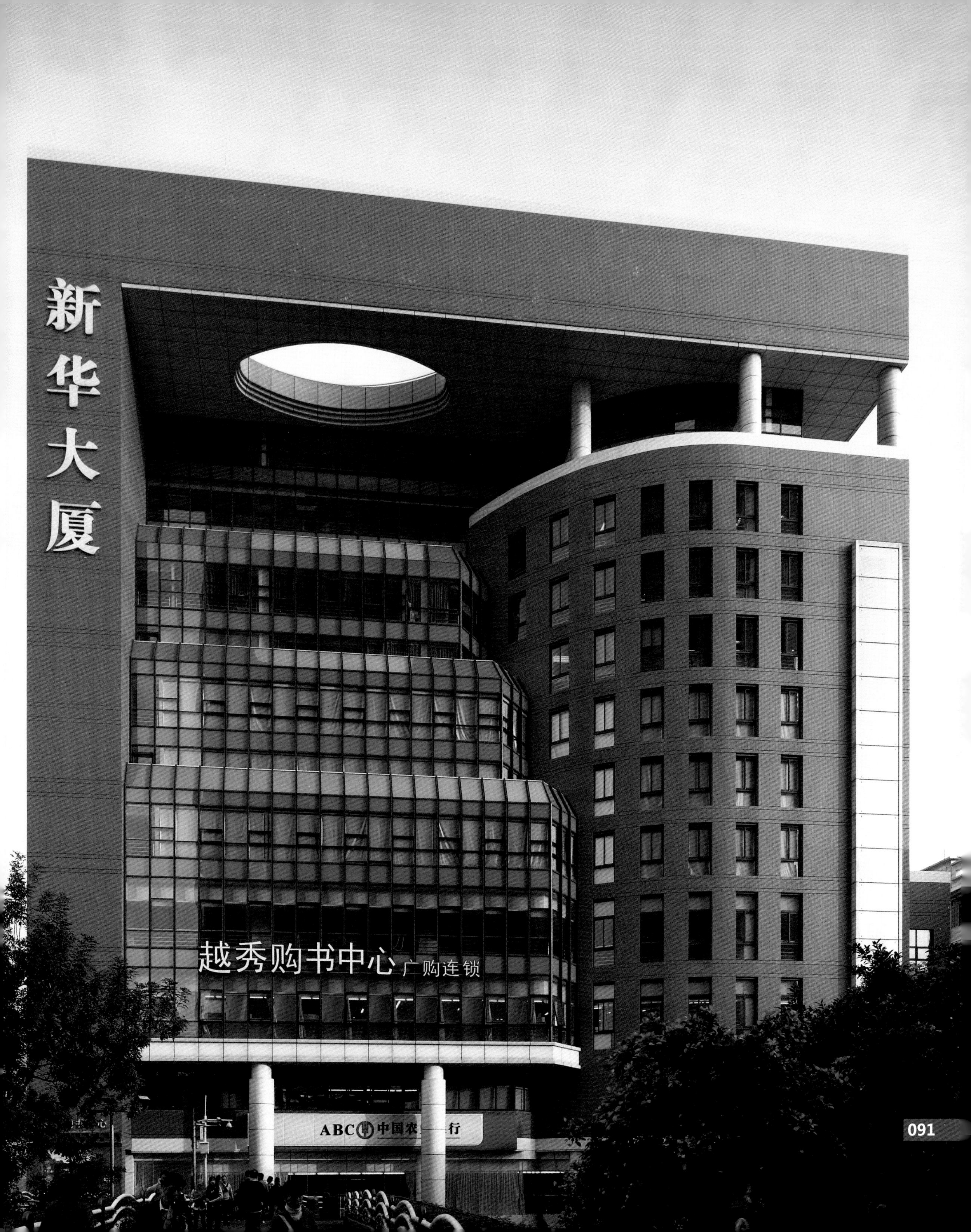
新华大厦
越秀购书中心
广购连锁
ABC

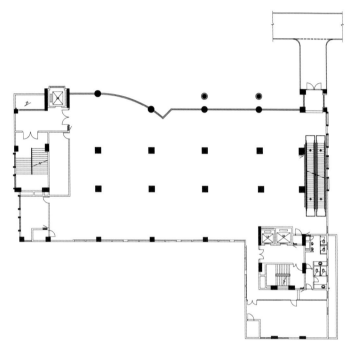

▲二层 *Level Two Plan*

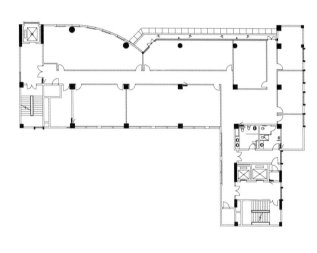

▲七层 *Level Seven Plan*

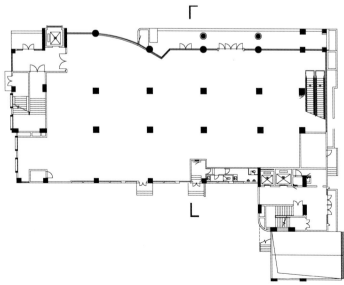

▲首层平面图 *Ground Level Plan*

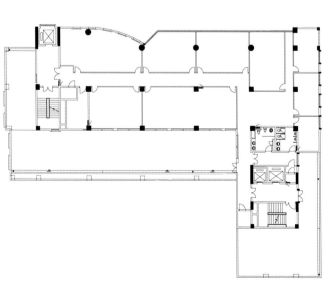

▲六层 *Level Six Plan*

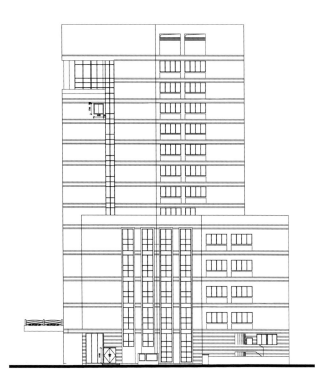

▲南立面图 South Elevation

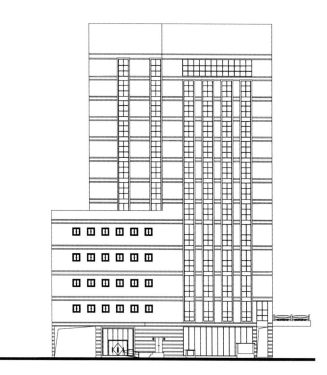

▲北立面图 North Elevation

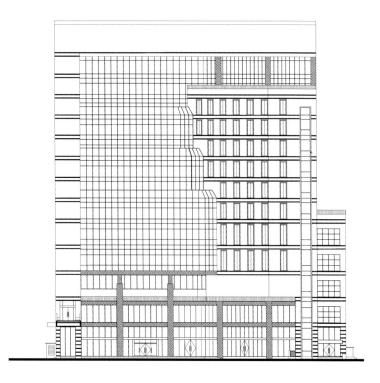

▲西立面图 West Elevation

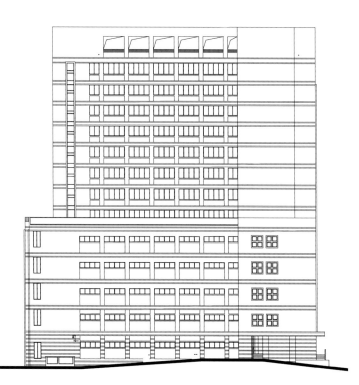

▲东立面图 East Elevation

佛山市世博商业中心
World Expo Business Center of Foshan

项目地点： Venue:	佛山市魁奇路 Kuiqi Road, Foshan
设计日期： Design Date:	2010 年 2010
项目建筑面积： Construction Area:	366,000 平方米 366,000m²
用地面积： Project Area:	86,822 平方米 86,822m²
曾获奖项： Awards:	设计竞赛中标 Bid Winner of Design Competition
合作设计： Partner :	黄军鹏、方良兵、黎就华、杨智敏 Huang Junpeng, Fang Liangbing, Li Jiuhua, Yang Zhimin

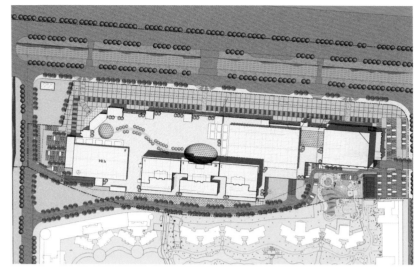

▲总平面图 *Master Plan*

佛山市世博商业中心是位于地铁上盖的佛山市超大型城市综合体，由大型商业中心、五星级酒店、公寓式办公楼组成，设计上注重创造动感新奇的购物空间，注重各种功能流线的安排，实现酒店、办公、商业最大化的资源共享，从而打造外形时尚，功能齐全的商业航母建筑。

World Expo Business Center of Foshan is a super-large complex situated above the metro station comprising of business center, five-star hotel and apartment office. Design focus on creating dynamic and fancy shopping environment and proper arrangement of various functional streamlines, so as to realize the maximum resource sharing of hotel, office and business utility, hence the flagship of business architecture is presented in contemporary shape and complete function.

▲模型图 *Building Model Plan*

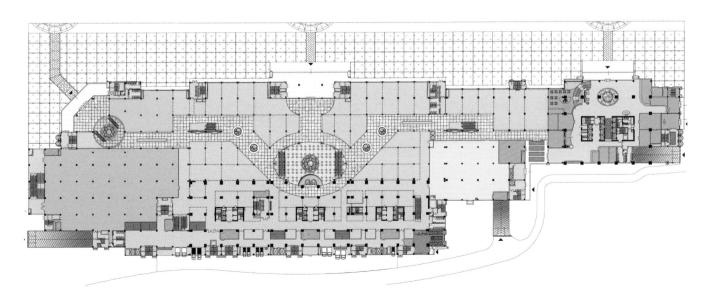

▲ 首层平面图 *Ground Level Plan*

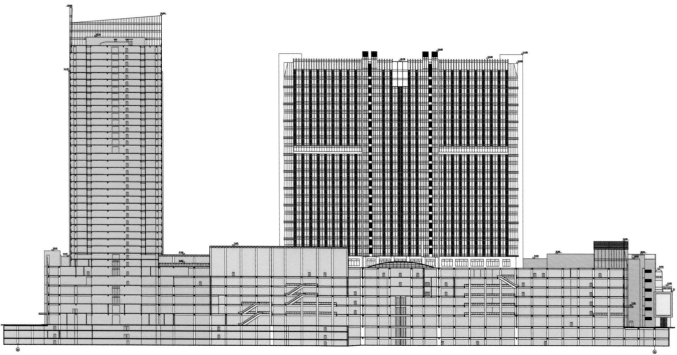

▲ 3-3 剖面图　3-3 Section

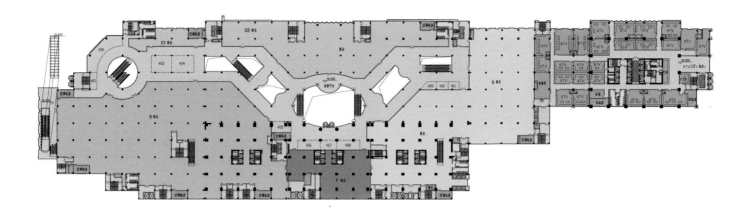

▲ 四层 *Level Four Plan*

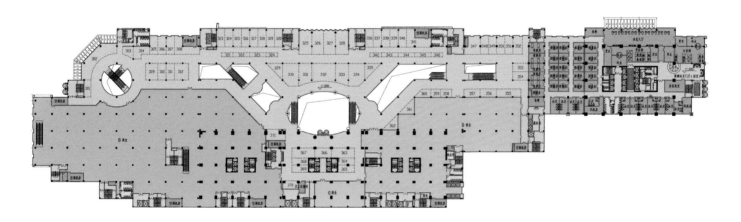

▲ 三层 *Level Three Plan*

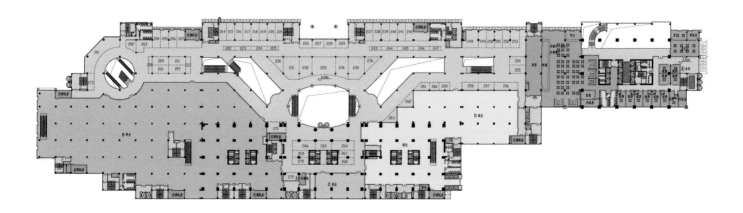

▲ 二层 *Level Two Plan*

佛山市南海区国土局地籍资料库
Cadastral Database of Land and Resources Bureau of Nanhai District of Foshan

项目地点： Venue:	佛山市南海区 Nanhai District, Foshan
设计日期： Design Date:	1999 年 1999
竣工日期： Completion Date:	2002 年 2002
项目建筑面积： Construction Area:	26,000 平方米 26,000m²
用地面积： Project Area:	9,800 平方米 9,800m²
曾获奖项： Awards:	广东省优秀建筑设计三等奖 Third Prize, Excellent Design of Guangdong Province
合作设计： Partner:	黄福生（方案），关小梅、候则林（施工图） Huang Fusheng(Project Design), Guan Xiaomei & Hou Zelin(Construction Drawing)

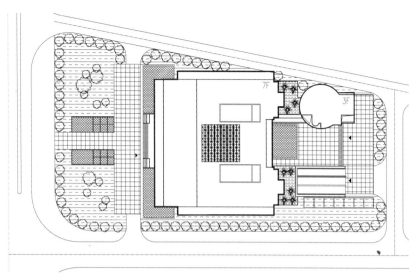

▲总平面图 *Master Plan*

佛山市南海区国土局地籍资料库在方形平面上采用"镂"的手法，镂出若干个形态各异的共享中庭，内部空间具有戏剧性的的丰富变化。建筑外形稳重大气，顶部大挑檐及柱托极具有中国古建筑的神韵，具有浓厚的文化特征，与大楼"国土管理"的使用性质相呼应。

The technique of "hollow-out" is employed on the square plan of Cadastral Database of Land and Resources Bureau of Nanhai District, several shared atriums are carved in various conformations, with dramatic and exuberant changes inside. The architecture is designed in steady and generous form, the large overhang eaves and column bracket imitating Chinese ancient architecture gives away luscious scent of culture, which echoes to the functional nature as "land management" of the Building.

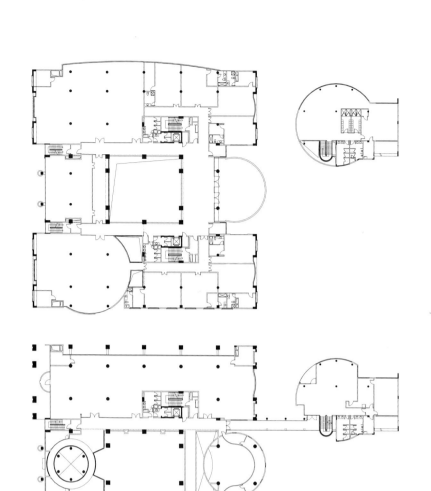

▲ 三层平面图 *Level Three Plan*

▲ 二层平面图 *Level Two Plan*

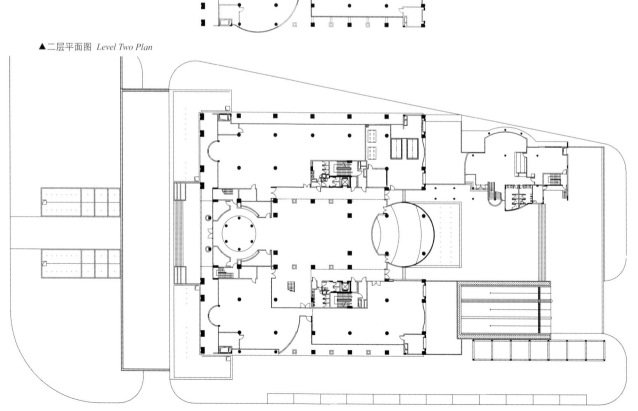

▲ 首层平面图 *Ground Level Plan*

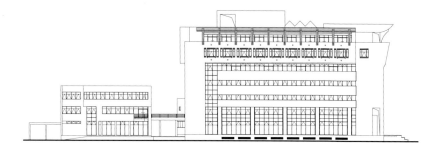

▲北立面图 *North Elevation*

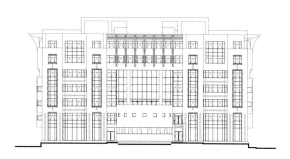

▲东立面图 *East Elevation*

▲南立面图 *South Elevation*

▲西立面图 *West Elevation*

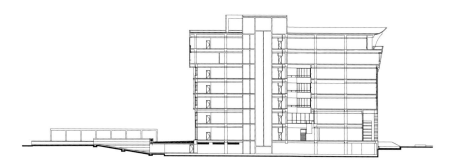

▲剖面图 *Section*

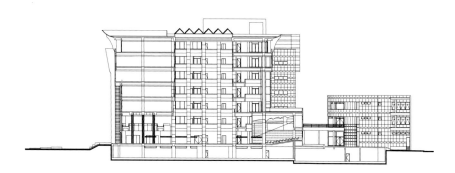

▲剖面图 *Section*

松岗广场
Songgang Square

项目地点： Venue:	佛山市南海区狮山镇松岗镇街 Songgang Street, Shishan Town, Nanhai District, Foshan
设计日期： Design Date:	2001 年 2001
竣工日期： Completion Date:	2009 年 2009
项目建筑面积： Construction Area:	14,127 平方米 14,127m²
曾获奖项： Awards:	① 住房和城乡建设部优秀建筑设计三等奖 ② 广东省优秀建筑设计二等奖 ① Third Prize, Excellent Design of Ministry of Construction ② Second Prize, Excellent Design of Guangdong Province
合作设计： Partner:	黄福生（方案）、黄考颖（施工图） Huang Fusheng(Project Design), Huang Kaoying(Construction Drawing)

▲总平面图 *Master Plan*

　　松岗广场的主体是一座生态型现代化行政办公大楼和一个市民广场，它座落于佛山市南海区松岗大道南侧，与松岗公园相邻，有小河蜿蜒其中，自然环境优美。

1. 天圆地方的总体布局
　　总平面采用天圆地方构图，由半圆形广场及中轴线作为衬托，建筑采用方形体形，座落于圆形水池之上，整体布局与公园、松岗河构图相协调，建筑与自然环境融为一体。

2. 生态节能的现代化办公环境
　　建筑平面以开敞式中庭为核心，周边布置办公空间，中庭四个方向均与空中花园及室外相连，所有办公室均有良好的自然采光通风，大楼不设中央空调，局部有特殊功能应设分体空调。

3. 建筑造型
　　大楼造型简约、明快，外形轻巧，具有新岭南主义的建筑特点，其屋顶深远的挑檐及墙面水平肋板造型美观，并有较好的遮阳及导风效果。柱廊式的造型与圆形水体互相衬托，刚柔并济，体现了天圆地方的中国文化理念。

　　大楼建成后，以其亲切的尺度、生态的办公环境，受到用户好评。

Songgang Square comprising an ecological contemporary administration office building and a Civic Center situates at the south side of Songgang Avenue of Nanhai. A river wriggles throughout the Square which is adjacent to the Songgang Park, the natural environment is advantaged and beautiful.

1. Master Plan of "dome-like heaven embraces the vast earth"
The master plan of "dome-like heaven embraces the vast earth" is complemented with semi-circular square and the central axis. The square-shape architecture sits on the circular pool. The plan is harmonious with the disposition of Songgang Park and Songgang River, perfectly integrated into the natural environment.

2. Ecological and energy-saving modern office
The open atrium is the core around which offices are placed. There are hanging gardens connecting to the outdoors in four directions of the atrium, hence all offices enjoy excellent natural light and ventilation. There is not central air-conditioning in the building, instead, separated air conditioners are installed for individual functional requirements.

3. Architecture Shape
The Building design is simple, clean and light full of Neo-Lingnan Style. The deep overhang eaves on the roof and horizontal floor on the wall are good-looking and as well serve for sun-shade and ventilation purpose. The colonnade works together with the circular waterscape to present the Chinese cultural philosophy of "dome-like heaven embraces the vast earth" in the form of couple hardness with softness.

The intimate and ecological Office Building has earned good praises from the users.

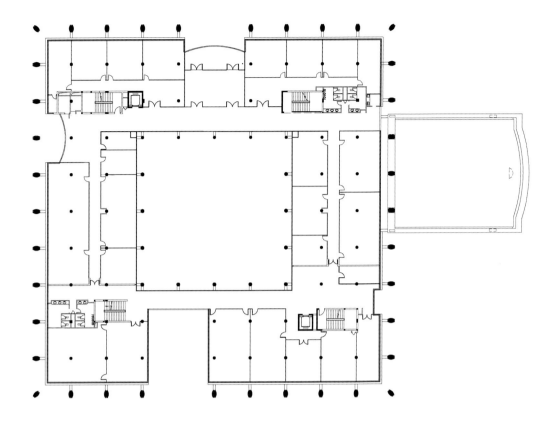

▲ 二层平面图 Level Two Plan

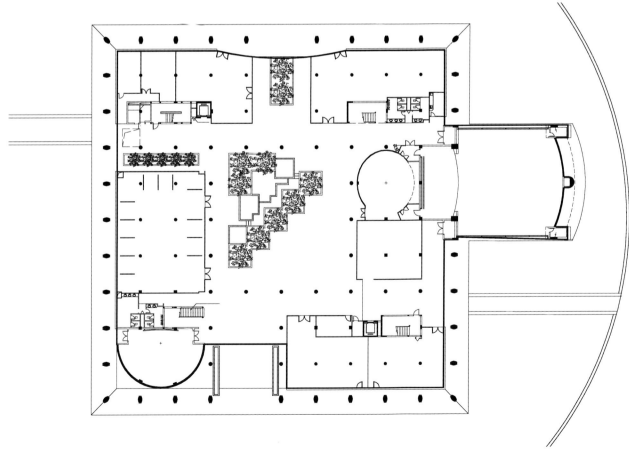

▲ 首层平面图 Ground Level Plan

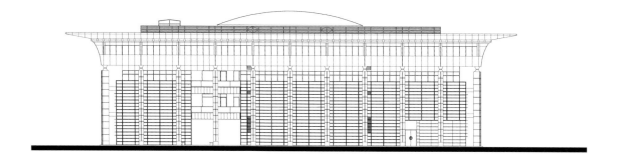
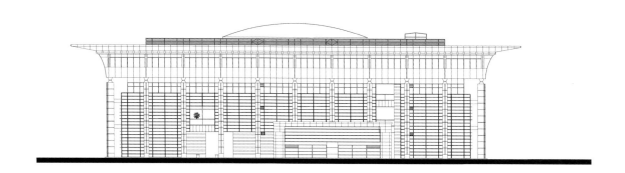
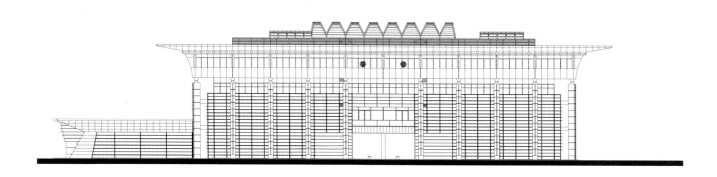
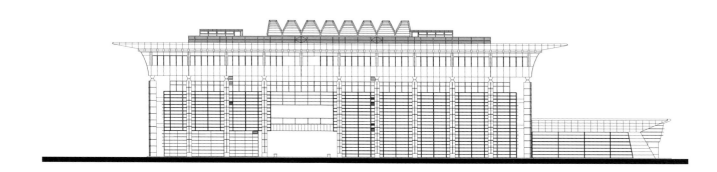

▲ 立面图 *Elevation*

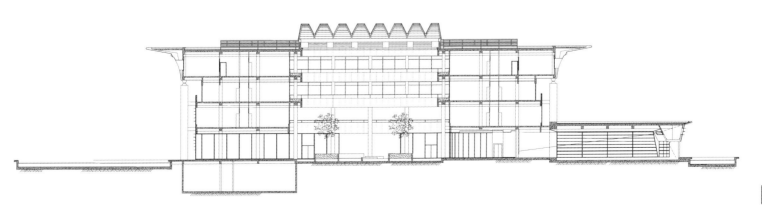

▲ 剖面图 Section

佛山市南海区国家税务局综合业务用房
Integrated Service Office of State Administration of Taxation of Nanhai District of Foshan

项目地点： Venue:	佛山市南海区 Nanhai District, Foshan
设计日期： Design Date:	2002 年 2002
竣工日期： Completion Date:	2008 年 2008
用地面积： Project Area:	16,135 平方米 16,135m²
项目建筑面积： Construction Area:	35,982 平方米 35,982m²
曾获奖项： Awards:	广东省优秀建筑设计二等奖 Second Prize, Excellent Architectural Design of Guangdong Province
合作设计： Partner:	黄福生（方案），黄考颖、黄惠青（施工图） Huang Fusheng(Project Design), Huang Kaoying & Huang Huiqing (Construction Drawing)

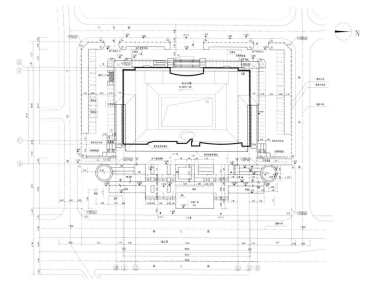

▲ 总平面图 *Master Plan*

佛山市南海国税综合业务用房是一个带有代表性、象征意义及文化价值的公共建筑物，它应脱离传统公共建筑物的庄严形象，它应是一栋对群众开放但具有雄伟、典雅、现代、亲切感的现代公共建筑物。它着重特有化建筑空间，使之成为大众能易于识别的标志；着重于内部空间的灵活使用，以适应现代社会对空间的需求及其改变的速度。因此，本项目突破传统办公室模式，把天然绿化环境延伸到办公空间，使之形成一个新理念的工作环境。

本项目为半围合式布局，人行主入口向东面市政主干道开放，车行入口面向西面小区路，周边道路形成环形消防车道。东面沿路设置大型绿化广场作为城市公共空间和建筑私密空间的过渡，为了方便市民的使用，建筑向市民开放的服务功能用房配置在绿化广场附近。东面公共绿化广场设有大台阶把客人直接引至建筑内庭花园，同时，湖滨景色也能一览无余。建筑的内外空间互相贯通，提高了建筑的透明性，强化了建筑与自然的融合关系。

Integrated Service Office of State Administration of Taxation of Nanhai District, Foshan is a significant and symbolic public architecture of cultural value which breaks away from the conventional solemn image. It appears to be a contemporary, grand, elegant and intimate public architecture open to the citizens. The unique architecture space makes it iconic to everyone, and the flexible usage of the internal space adapts it to the space demand and changeful rhythm of modern society. Hence the design breaks through the traditional office pattern, extends the natural green environment to the office and brings forth the working ambience of new concept.

Then semi-enclosed layout sets the pedestrian main entrance open to the eastern key street, the vehicle entrance faces to the western path of the community, and surrounding roads form the annular fire traffic lane. Large green plaza is designed along the way in the east as the transition from public city space to private architecture void. For the citizens' convenience, the service functional rooms open to the citizens are placed next to the green plaza. There are large steps in the eastern public green plaza leading the visitors to the inner garden; meanwhile the beautiful lakeshore view can be taken in a glance. The interior and external space is connected to each other to enhance the transparency of the architecture, as well to reinforce the harmony between architecture and nature.

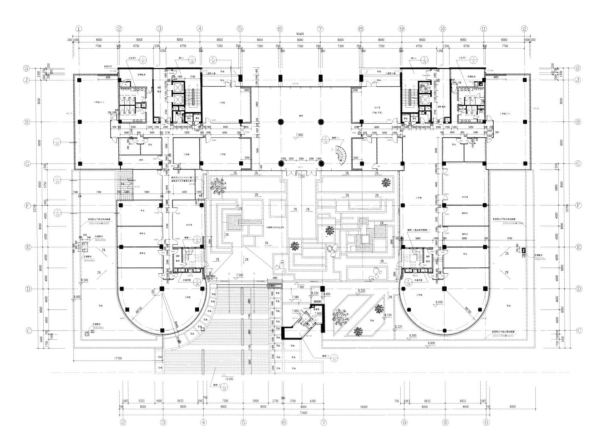

▲二层平面图 *Level Two Plan*

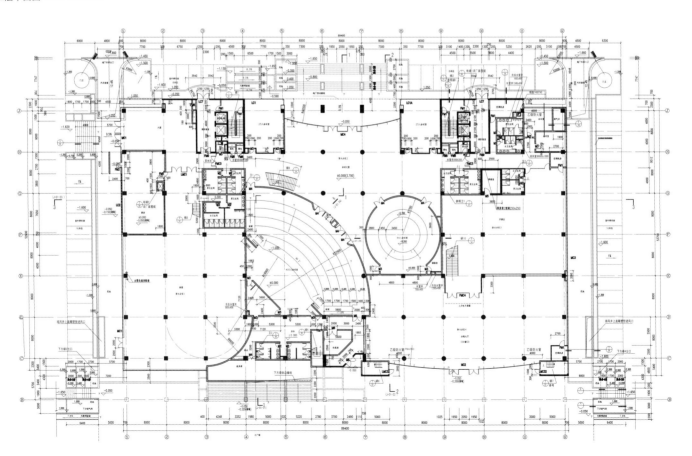

▲首层平面图 *Ground Level Plan*

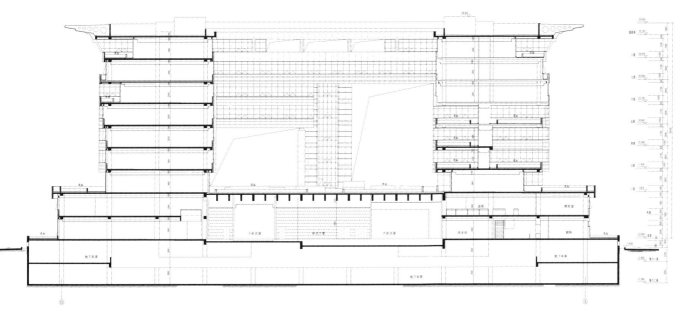

▲ 剖面图 Section

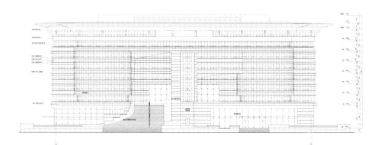

▲ 东立面图 *East Elevation*

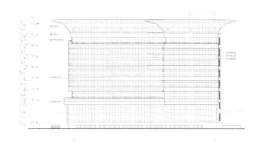

▲ 北立面图 *North Elevation*

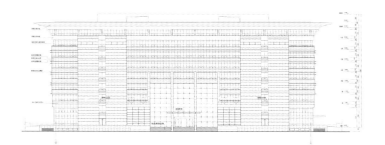

▲ 西立面图 *West Elevation*

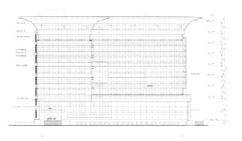

▲ 南立面图 *South Elevation*

南宁市云星办公楼
Yunxing Office Building of Nanning

项目地点： Venue:	广西南宁 Nanning, Guangxi Province
设计日期： Design Date:	2010 年 2010
竣工日期： Completion Date:	2012 年 2012
项目建筑面积： Construction Area:	20,131 平方米 20,131m²
用地面积： Project Area:	2,470 平方米 2,470m²
合作设计： Partner :	颜立华 Yan Lihua

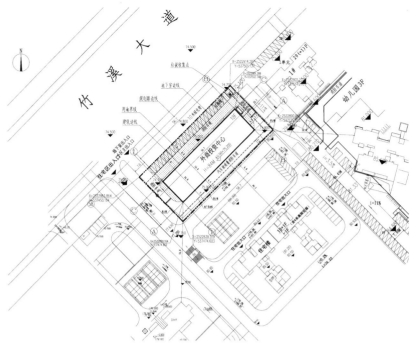

▲总平面图 *Master Plan*

南宁市云星办公楼项目倡导"绿色办公24小时"的建筑概念，办公室可灵活采用多种空调形式，可满足全天候办公的要求，其立面设计以竖线条为主，框线有精致的细部设计，造型简约纯净，体现了"少就是多"的设计理念。

Aiming to advocate the architecture concept of "around-the-clock green office", Yunxing Office Building of Nanning flexibly combines various air-conditioning to fulfill the all-day office requirement. Vertical line dominates the elevation design, and the delicate detail design of the border express the idea of "less is more" in the simple and pure shape.

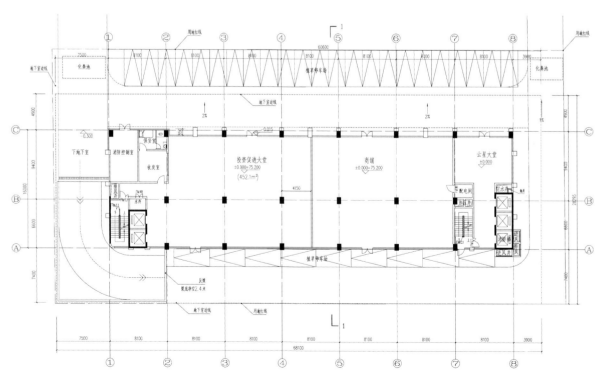

▲首层平面图 *Ground Level Plan*

121

惠州市惠东博物馆
Huidong County Museum of Huizhou

项目地点： Venue:	广东省惠州市惠东 Huidong, Huizhou, Guangdong
设计日期： Design Date:	2013 年 2013
竣工日期： Completion Date:	2014 年 2014
项目建筑面积： Construction Area:	7,500 平方米 7,500m²

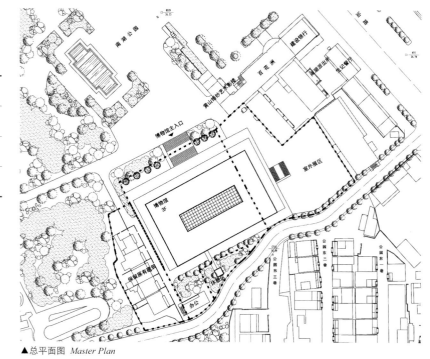

▲总平面图 *Master Plan*

惠东博物馆是在原有地块的基础上，利用东边的闲置用地进行建设。拟建博物馆高三层，以红色宝盒和水立方的造型组合，体现惠东县特有的红色文化和海洋文化，强调形体构成与虚实结合，与周边环境协调统一。外墙使用红色砂岩，室外广场采用卵石铺贴，均为当地特有建筑材料。

Huidong County Museum is built on the base of original plot utilizing the unused land in the east. The Museum is designed as three stories, in the shape of red treasure box and water cube combination, to display the unique red culture and marine culture of Huidong County. It focuses on the form and structure composition, false and true combination, as well coordination and unity with surrounding environment. The red sandstone of exterior walls and pebble flooring of the outdoor square are both indigenous construction materials.

▲黄劲手绘效果图 *Hand drawing, Huang Jin*

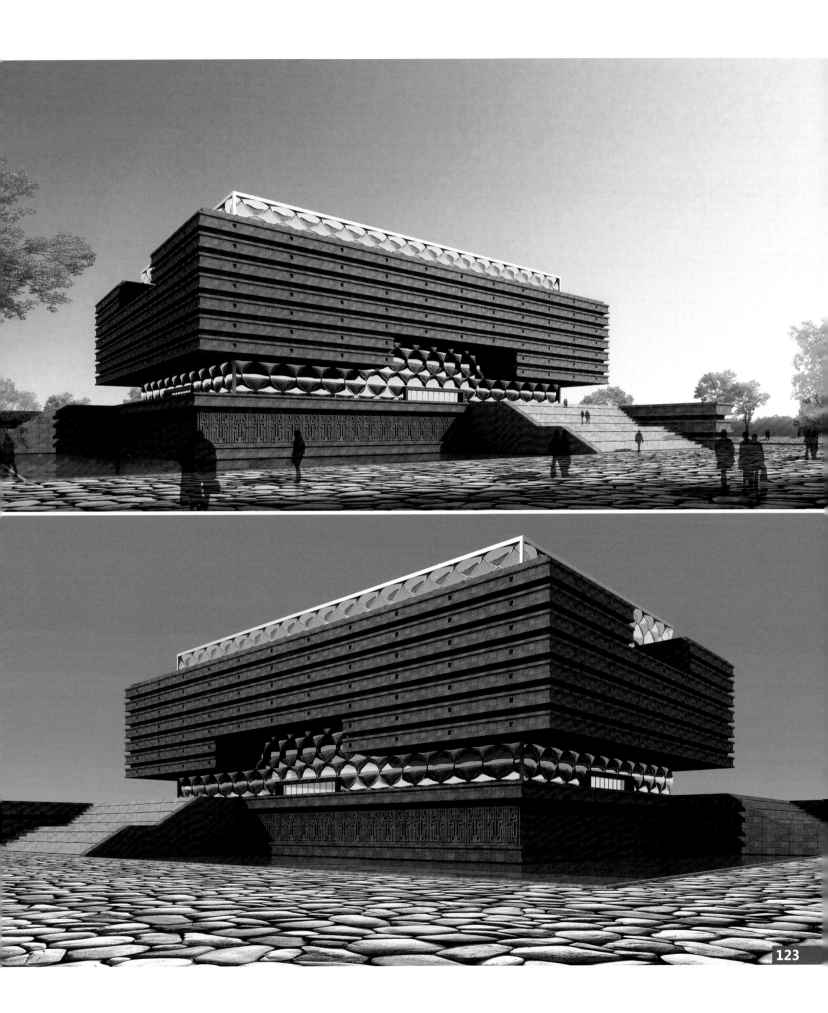

北海市银滩风帆大酒店
Silver Beach & Sail Hotel, Beihai

广州云来斯堡酒店、办公综合楼
Vanburgh Hotel, Office Building

南昌市新好景大酒店、别墅区
Hotel & Villa of Xinhaojing Hotel, Nanchang

酒店建筑 HOSPITALITY

海南省三亚市福朋喜来登大酒店
Four Points by Sheraton Sanya, Hainan Province

老挝欧亚首脑峰会接待大酒店
Landmark Meckong Riverside Hotel Architecture & Interior design of Asia-Europe Meeting Summit Hotel, Laos

老挝万象广晟大酒店
Guangsheng Hotel, Vientiane, Laos

佛山市南国桃园枫丹白鹭酒店
Fontainebleau Hotel, Nanguo Peach Garden, Foshan

鼎湖·总统御山莊
Ding Hu·Presidential Royal Villas

中山市翠景大酒店及公寓
Cui Jing Hotel & Apartment, Zhongshan

哈尔滨东方红旅游区温泉酒店
Harbin Dongfanghong Tourist Area Hot Springs Hotel

福州马尾海峡旅游区酒店
Fuzhou Mawei Strait Tourist Area Hotel

霸王大酒店
Bawang hotel

北海市银滩风帆大酒店
Silver Beach & Sail Hotel, Beihai

项目地点： Venue:	广西北海市 Beihai, Guangxi Province
设计日期： Design Date:	2009 年 2009
竣工日期： Completion Date:	设计中 Design Stage
用地面积： Preject Area:	7,260 平方米 7,260m²
项目建筑面积： Construction Area:	50,000 平方米 50,000m²
合作设计： Partner :	颜立华 Yan Lihua

银滩风帆大酒店南邻著名的北海银滩，是生态型的五星级度假酒店，酒店塔楼平面为L形，采用单廊式布置，所有客房均可观海，均可享受天然的通风和阳光，其外形以风帆为寓意，两扇高低错落的帆形造型极具标志性。立面的景观玻璃阳台造型丰富，多变的弧形阳台犹如海上波浪，在阳台照耀下高低起伏，极具浪漫情调。

Silver Beach & Sail Hotel is an ecological five-star resort hotel adjoining the famous Beihai Silver Beach. The "L" shape tower building applies peripteral arrangement that ensure good seaview, natural ventilation and sunshine in all guest rooms. The sail-outline originating from the design with two well-proportioned sails forms symbolic architecture. The landscape glass balcony is designed in various shapes. The changing curvy balconies undulate under the sun in a romantic mood as if the waves of Beihai.

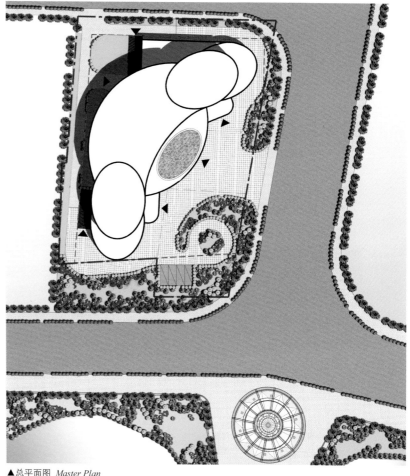

▲总平面图 *Master Plan*

▲黄劲手绘效果图 *Hand drawing, Huang Jin*

▲ 细部效果图 *Rendering of Detail*

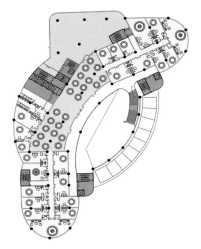

▲ 二层平面图 *Level Two Plan*

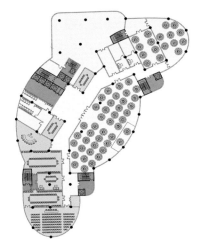

▲ 三层平面图 *Level Three Plan*

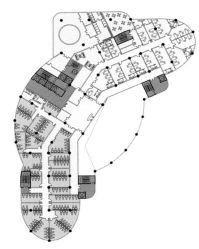

▲ 四层平面图 *Level Four Plan*

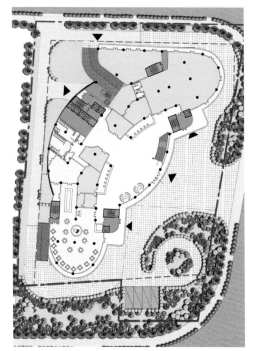

▲ 首层平面图 *Ground Level Plan*

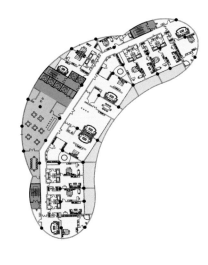

▲ 行政楼标准层平面图

Standard Floor Plan of Guest Room

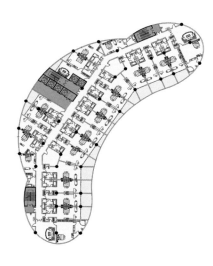

▲ 客房标准层平面图

Standard Floor Plan of Apartment

海南省三亚市福朋喜来登大酒店
Four Points by Sheraton Sanya, Hainan Province

项目地点： Venue:	海南省三亚市三亚湾路 78 号 No.78 Sanya Bay Road, Sanya Bay, Hainan
设计日期： Design Date:	2007 年 2007
竣工日期： Completion Date:	2012 年 2012
项目建筑面积： Construction Area:	35,000 平方米 35,000m²
合作者： Partner:	巴马丹拿设计事务所，黄考颖、候则林、蔡礼帮、洪琰 Palmer & Turner Group, P&T, Huang Kaoying, Hou Zelin, Cai Libang, Hong Yan

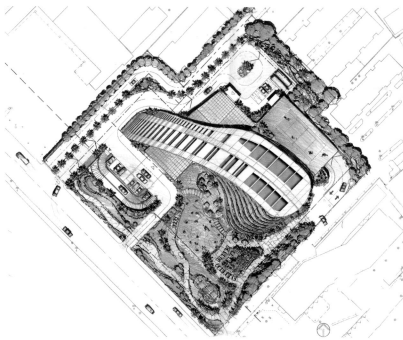

▲ 总平面图 *Master Plan*

福朋喜来登大酒店是一座时尚的海滨城市酒店，设计上将主入口设于背面二层，以获取良好的室外景观，宾客在二层大堂即可俯瞰三亚湾的优美风光。塔楼后勤用房设于背面，正面全部客房客均可观赏海景。其立面的风帆造型极具动感，旋转上升的造型在不同角度均有独特的形象，成为三亚湾新的标志建筑。

Four Points by Sheraton is a fashionable seashore hotel in Sanya Bay. The entrance is designed on the second floor at the back of the building to ensure wonderful outdoor scenery, where the guests are able to overlook the beautiful view of Sanya Bay right from the lobby. The logistics rooms are arranged at the back of the tower building, and all guest rooms in the front could enjoy sea view. The sail-shape hotel appears to be so dynamic with unique looks during circumrotating and rising from different angles. Four Points by Sheraton has become the new iconic architecture of Sanya Bay.

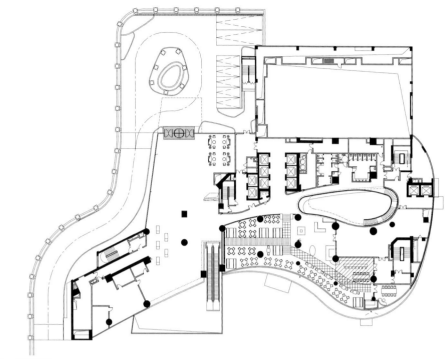

▲ 二层平面图 *Level Two Plan*

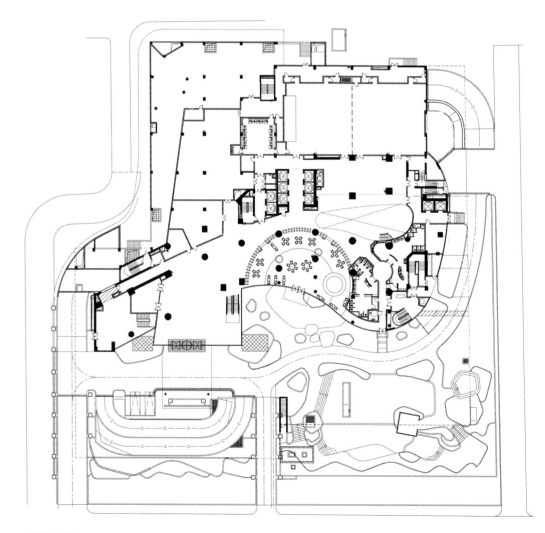

▲ 首层平面图 *Ground Level Plan*

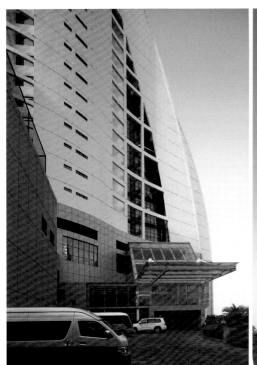

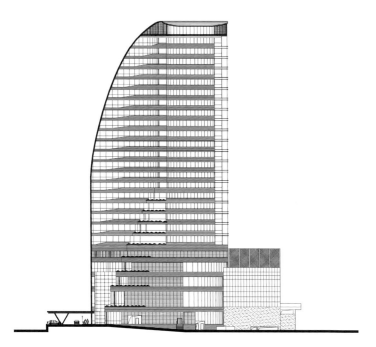

▲ 东立面图 *East Elevation*

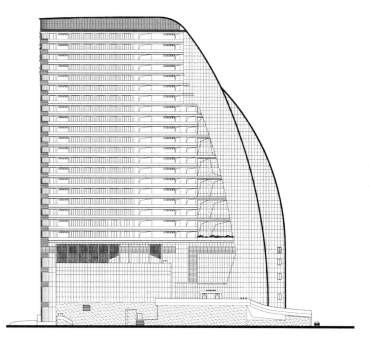

▲ 北立面图 *North Elevation*

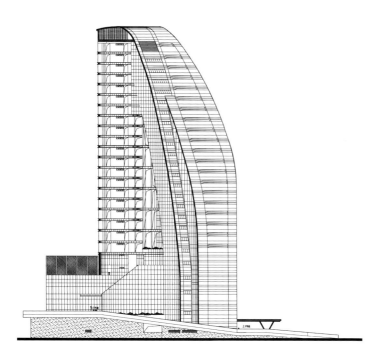

▲ 西立面图 *West Elevation*

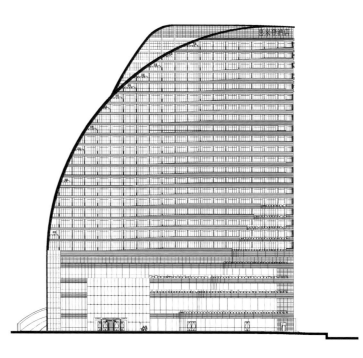

▲ 南立面图 *South Elevation*

老挝亚欧首脑峰会接待大酒店
Landmark Meckong Riverside Hotel

Landmark Meckong Riverside Hotel of Asia-Europe Meeting Summit Hotel, Laos

项目地点： Venue:	老挝万象 Vientiane, Laos
设计日期： Design Date:	2012 年 2012
竣工日期： Completion Date:	2012 年 2012
用地面积： Project Area:	50,000 平方米 50,000m²
项目建筑面积： Construction Area:	35,000 平方米 35,000m²
曾获奖项： Award:	国际竞赛中标 Bid Winner of International Design Competition
合作设计： Partner:	颜立华、杨智敏、朱永东、郭文河 Yan Lihua, Yang Zhimin, Zhu Yongdong, Guo Wenhe

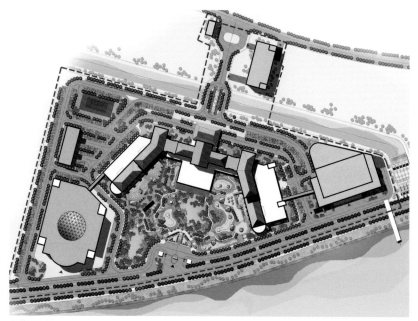

▲ 总平面图 *Master Plan*

　　老挝欧亚首脑峰会接待大酒店是为迎接峰会51个国家及地区的总统、首脑要员及随员，由吉达蓬公司投资兴建的老挝特色五星级城市度假酒店。

　　"景观优先"是本酒店设计理念。酒店坐落于美丽的湄公河畔，平面采用单廊式设计，围绕庭园布置，所有房间均可观赏河景。

　　酒店大堂设于二层，可俯瞰湄公河，其大堂空间作为重点设计，从门廊、大堂、咖啡厅到室外观光平台，空间极具趣味性，且层次变化丰富。

　　老挝是个多民族、多文化的国家，设计动笔前设计师遍访老挝古今建筑，搜集其建筑精华，最终所选用的建筑风格体现了老挝"多元文化"的特点，即以老挝当地的建筑风格为主，并融入法式、中式等外来建筑文化。

　　酒店建筑立面采用三段式，裙房为"干阑式"（干栏式）穹廊结构，具有法式建筑及当地建筑特点，顶部造型层层跌级，轮廓线十分丰富，坡顶采用了老挝歇山建筑形式，正中坡顶借鉴了老挝皇宫的造型（世界物质遗产），给建筑增添了不少历史文化元素。

Asia-Europe Meeting Summit Hotel of Laos is a five-star resort hotel of Laos style invested by Krittaphong Group, it is constructed to accommodate the Presidents, Leaders and attendees for the Summit from 51 countries.

"Landscape first" being the design philosophy, the hotel sits by the beautiful River Mekong. The peripteral design around the courtyard enables all rooms access to the excellent river view.

The Lobby is set on the second floor in focal design with a panoramic view of River Mekong. The space is full of interest and layer changes from outdoor sightseeing platform.

Laos is a melting pot of different nationalities and cultures. The designer visited various traditional and contemporary architectures to derive the essence. The final design comes out to be the one that best express the feature of the "diversified culture" of Laos, which focus on the local architecture style of Laos while infusing exotic culture and style such as French and Chinese.

The elevation of the hotel is divided into three sections—the podium introduces stilted arc-top corridor, a fusion of French and local architecture style. The descending levels of steps at the top delineate affluent silhouette; the slope top follows gable and hip roof architecture style of Laos; the middle part is enlightened by the form of Lao Palace (World Heritage), adding considerate historical and cultural elements to the architecture.

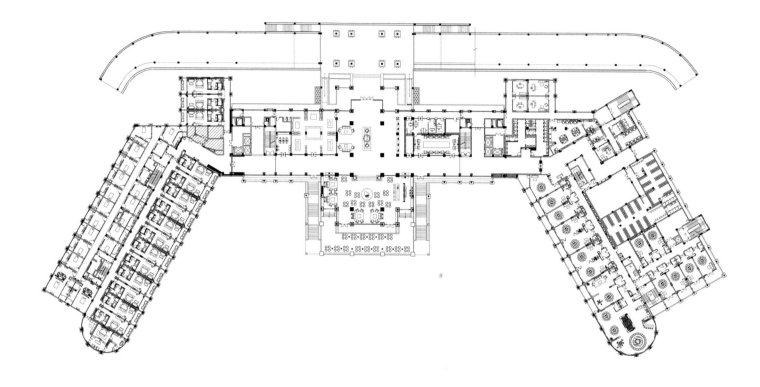

▲一层平面图 *Level One Plan*

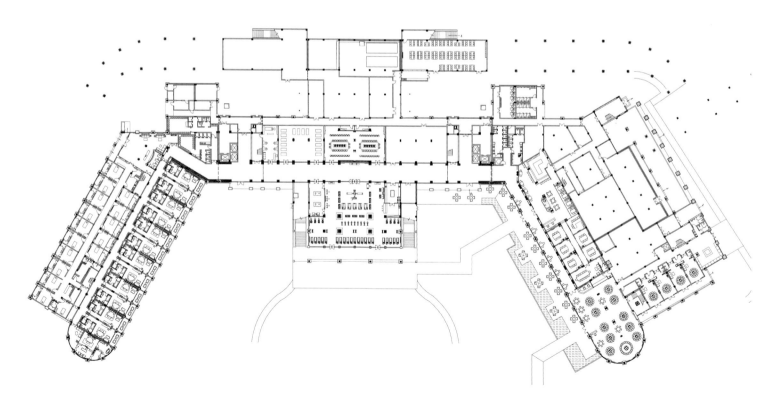

▲地面层平面图 *Ground Level Plan*

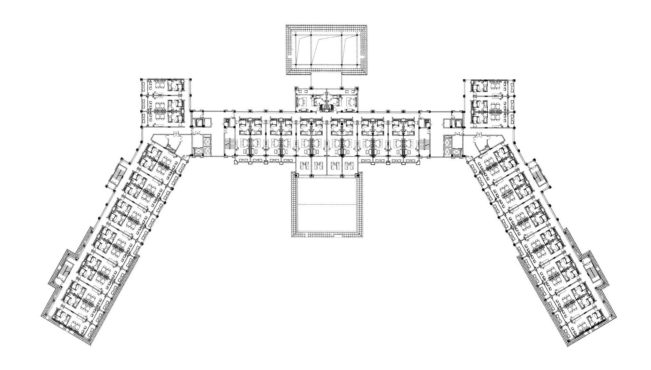

▲ 三层平面图 *Level Three Plan*

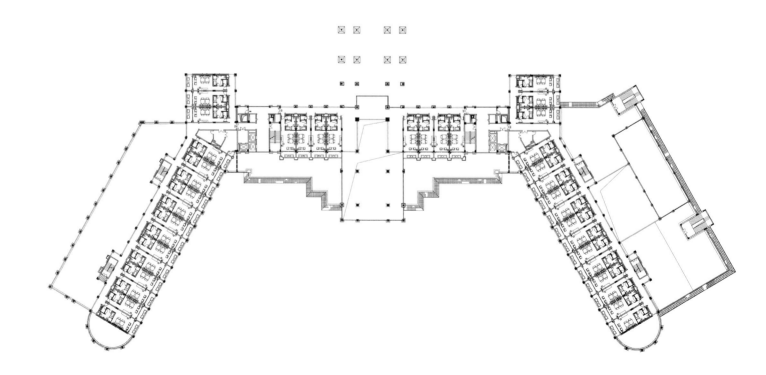

▲ 二层平面图 *Level Two Plan*

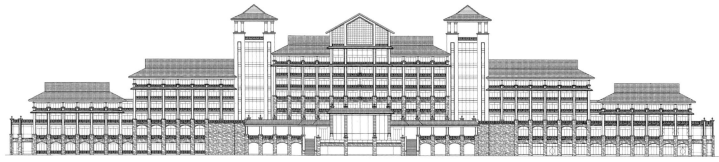
▲东立面图 *East Elevation*

▲西立面图 *West Elevation*

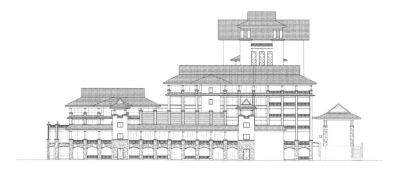
▲南立面图 *South Elevation*

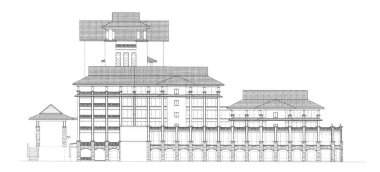
▲北立面图 *North Elevation*

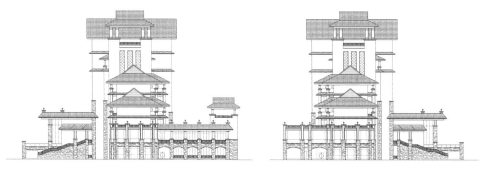
▲西南和西北立面图 *W-S & W-N Elevation*

老挝万象广晟大酒店
Guangsheng Hotel, Vientiane, Laos

项目地点： Venue:	老挝万象市塔峦 That luang, Vientiane, laos
设计日期： Design Date:	2012 年 2012
竣工日期： Completion Date:	2013 年 2013
用地面积： Project Area:	14,377 平方米 14,377m²
项目建筑面积： Construction Area:	14,400 平方米 14,400m²
曾获奖项： Awards:	国际竞赛中标 Bid Winner of International Design Competition

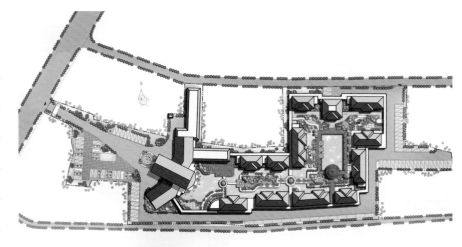

▲ 总平面图 *Master Plan*

老挝万象广晟大酒店定位为园林式精品酒店，以体现老挝的特色，小而全的配套为经营特点。酒店采用多层园林式的布局设计，所有房间均可观赏内庭园林，建筑借鉴了当地"干阑式"（干栏式）建筑形式，首层架空用作车库，适应当地炎热多雨气候。建筑风格颇为独特，融合了老挝和法式的建筑文化特点。

酒店竖向采用"微坡地"设计，入口处利用架空层升起的高度升高3m，内庭园利用高差形成微坡地高低变化，园林空间层次颇为丰富。

Guangsheng Hotel in Vientiane of Laos is a garden boutique hotel. "Small but complete" becomes the feature of the design, which also presents the trait of Laos. The layout design of layered garden is adopted to enable good view of the inner courtyard from all guest rooms. The architecture design is inspired by the local "Stilt Style", the first floor is stilted as the car park, which adapts to the torrid and humid climate. The unique building is a combination of Laos and French architectural and cultural styles.

Vertically "gentle slope" design is employed. The height of the entrance is increased three meters by the stilted level. The inner courtyard shows high and low variations of gentle slopes due to height difference, hence creates nice garden of rich layers.

▲ 酒店立面图 *Elevation of Hotel*

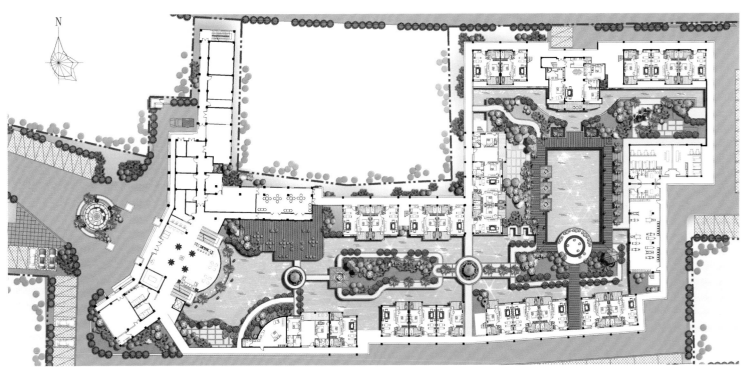

▲ 酒店首层平面图 *Ground Level Plan of Hotel*

广州云来斯堡酒店、办公综合楼
Vanburgh Hotel, Office Building

项目地点： Venue:	广州市珠江新城 Zhujiang New Town, Guangzhou
设计日期： Design Date:	2007 年 2007
竣工日期： Completion Date:	2012 年 2012
项目建筑面积： Construction Area:	89,700 平方米 89,700m²
用地面积： Project Area:	13,215 平方米 13,215m²
曾获奖项： Awards:	①设计竞赛中标 ② 2014 年度全国工程建设项目优秀设计成果二等奖 ① Bid Winner of Design ② Second Prize, 2014 annual national construction project design excellence awarded
合作设计： Partner:	方良兵、颜立华、黄国明 Fang Liangbing, Yan Lihua, Huang Guoming

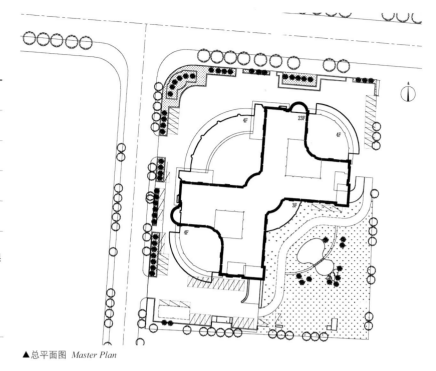

▲ 总平面图 *Master Plan*

具有浪漫气质的五星级云来斯堡酒店已成为广州核心城区东入口处的标志性建筑，其塔楼由客房和办公楼连体构成十字形平面，在用地四个角位留出广场空间，分别给酒店、办公、餐饮、娱乐独立使用，功能流线清晰合理。十字形平面在有限的用地上划出宽阔的入口广场和后花园，为大堂及咖啡厅创造良好的室外景观。外形采用"各向主立面"设计，十字形平面布置带来形象均好的四个立面，裙房给不同功能用户创造良好的入口形象，建筑造型方圆结合，形体和细部经过推敲，色彩柔和，造型优雅。

酒店以具有中国特色的"祥云"作为文化主题，贯穿于酒店建筑室内设计之中，酒店大堂云形天花、宴会厅折扇造型等设计构思，将中国古典文化与时尚生活巧妙地融合于一体。

The romantic five-star Vanburgh Hotel is the iconic architecture at the east entrance of the downtown core of Guangzhou. The hotel tower is in the shape of crucifix composed of guest house building and office building. The four corners are retained for the square to accommodate hotel, office, restaurant and entertainment spaces, which are functionally designed in a clean and reasonable way. The "crucifix" delineates a spacious entrance square and backyard garden on the limited land, creating wonderful outdoor view for the lobby as well as the coffee bar. The form employs "west-oriented elevation" design. The layout of crucifix engenders four nice elevations whilst the podiums set up an impressive entrance image for various functional users. The architecture is perfectly integrated with square and circle in meticulous forms and details, presenting elegant shape in mild colors.

"Auspicious Cloud" of Chinese feature becomes the cultural theme of the hotel and impenetrates the entire interior design. Chinese classical

▲ 黄劲手绘效果图 *Hand drawing, Huang Jin*

culture and contemporary living are artfully merged by design concepts such as cloud-shape ceiling in the lobby, and folding fan patterns in the banqueting hall.

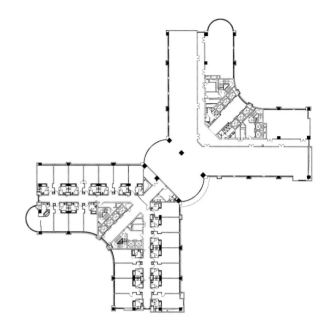

▲ 六至九层平面图 *Level Six-nine Plan*

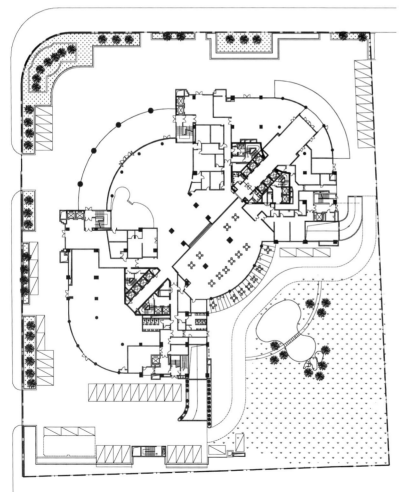

▲ 首层平面图 *Ground Level Plan*

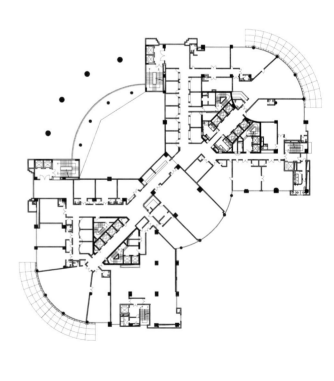

▲ 二层平面图 *Level Two Plan*

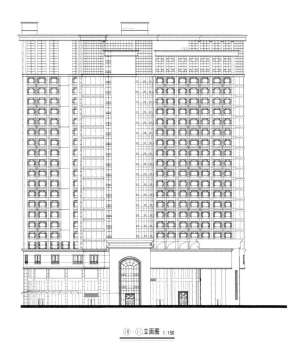

▲北立面图 North Elevation

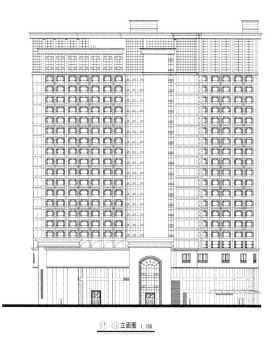

▲西北立面图 Northwest Elevation

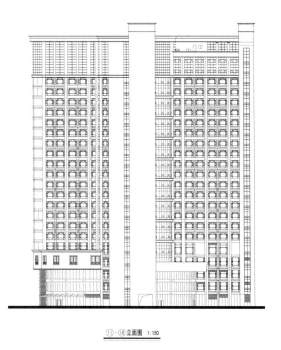

▲南立面图 South Elevation

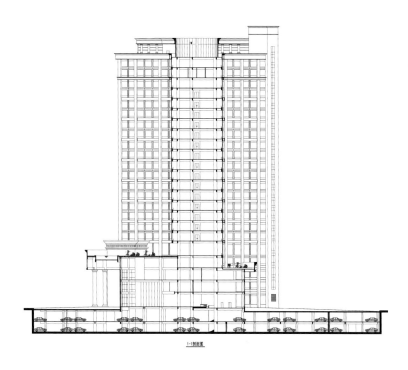

▲剖面图 Section

佛山市南国桃园枫丹白鹭酒店
Fontainebleau Hotel, Nanguo Peach Garden, Foshan

项目地点： Venue:	广东省佛山市南国桃园度假区 Nanguo Peach Garden Tour Resort in Nanhai District, Foshan, Guangdong
设计日期： Design Date:	1997年 1997
竣工日期： Completion Date:	1999年 1999
项目建筑面积： Construction Area:	45,000平方米 45,000m²
曾获奖项： Awards:	广东省优秀建筑设计二等奖 Second Prize, Excellent Architectural Design of Guangdong Province
合作设计： Partner:	黄福生（方案），黄考颖、黄惠青、周茂（施工图） Huang Fusheng(Project Design), Huang Kaoying & Huang Huiqing & Zhou Mao(Construction Drawing)

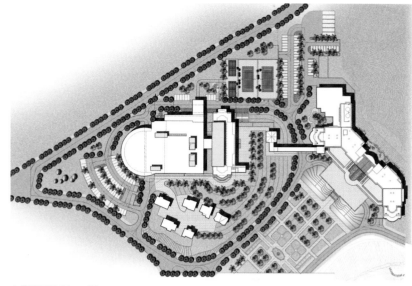

▲总平面图 *Master Plan*

佛山市南国枫丹白鹭酒店位于佛山市南海区南国桃园旅游区内，占地7万平方米，经营以度假、会议为主。当地原有侨乡的中西合璧风格的建筑激发了项目创作的灵感，酒店建筑采用了简洁的欧陆建筑风格。酒店顺应其东、西、北三面环山，南面靠湖的有利地形进行总体布置，从北面山丘至湖边，依次布置酒店及三个主题公园，三个公园从北至南层层跌级，依次展开艺术图案、绿化图案、观景台、艺术雕塑、跌级流水、花架长廊和喷泉广场等景观，这布局规则、轴线对称及整体造园风格的人工化的欧式传统园林为酒店创造了十分优美的园林环境。

酒店主楼的建筑面积为2万平方米，地下层为车库、设备房；首层为大堂、中西餐厅、180人会议大厅、健康中心和泳池；二层为中小型会议室及客房，三至六层为客房；主楼共有230间客房。

主楼空间设计注重引入室外环境及创造良好的视觉空间，大堂、酒吧、餐厅、客房均有不同的室外景观。二层高的大堂比室外地坪升高2米，向南可俯视公园全景，向北可观赏山边的叠石瀑布。大堂设计以暖色调为主，具有温馨浪漫的气氛，弧形楼梯及白鹭铜雕给大堂增添了不少艺术气氛。

主楼平面呈折线形变化，体形舒展开朗，从上而下呈跌级变化，与山势相呼应，同时显得活泼自然。外形上兼具欧陆情调和现代气息。檐口、墙面、阳台等均有较为精致的细部处理。

国际会议中心位于酒店西南部的山丘上，包括会议中心、停车库、网球场等项目，总用地约2万平方米。国际会议中心大楼由会议中心和客房楼组成，分别列于大楼东、西两翼，中

Fontainebleau Hotel is located within Nanguo Peach Garden Resort of Foshan City with a total area of 70,000m². Resort and commercial meetings are two major business of the Hotel. The hometown of overseas Chinese building style-to combine the eastern and western architectural style-inspired the project inspiration. The Hotel is surrounded by mountains on its east, west and north side, and lake to the south. The design is in consistence with the advantageous landform, arranged hotel and three theme parks from northern hill to the lake. Three parks are with stepped distribution characteristics from north to south. Landscape are sequentially deployed of artistic planting patterns, viewing platform, aesthetic sculptures, dropping cascades, flower corridor and fountain square etc.. A beautiful garden environment is created for the hotel with the artificial traditional garden in European style of exquisite layout plan, symmetrical axis and uniform approach.

The main building's construction area of 20,000m², in basement there are car park and equipment rooms; the First Floor accommodates the lobby, Chinese & Western Restaurants, Conference Hall for 180 people, Fitness Center and Swimming Pool; the Second Floor is allocated for Middle & Small Conference Rooms and Guest Rooms; the Third Floor to the Six Floor are all Guest Rooms.

The space design focus on the introducing outdoor environment and creating excellent visual effect, hence there are different exterior landscape for lobby, bar, restaurants and guest rooms. The two storey-high lobby is two meters higher than outside, from where the entire park to the south could be overlooked, and rockery setting and waterfall view to the north could be enjoyed. Warm color dominating the lobby

▲ 立面图 *Elevation*

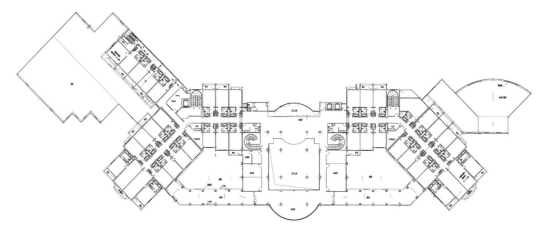

▲ 主楼二层平面图 *Level Two Plan*

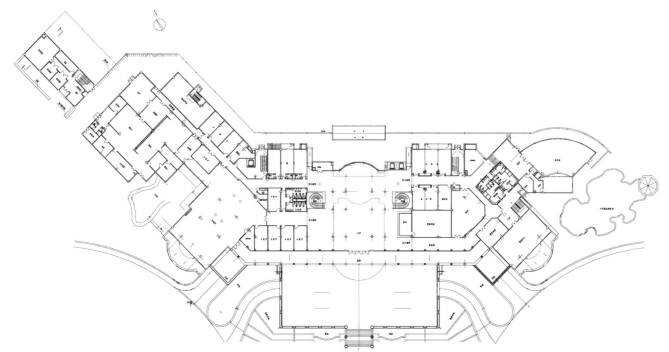

▲ 主楼首层平面图 *Ground Level Plan*

央以庭园相连，大楼与枫丹白鹭酒店的主楼由东侧的一条连廊连成一体。建筑依山而建，利用地形的高差设置了半地下层，布置后勤用房和设备房，节省了投资。会议中心及客房楼分设出入口，人流互不干扰。

平面布局设计上，会议中心首层设 1 个 100 座的会议厅，2 个 50 座的会议厅，1 个多功能演艺厅及 8 个小型会议室，二层为大型宴会厅；客房楼首层为大堂，二至六层为客房，屋面均作绿化设计，有效地保温隔热。

立面设计上，大楼以方形体型为主，西侧路口方向呈半圆形布置，居于高处以通透玻璃将四周山林景色引入室内，东侧以实面为主，建筑细部注重比例及细部处理，以干挂石为主的具雕塑感的造型及精致的细部处理，体现其亲切的形象及体贴入微的个性。

design infuses the atmosphere filled with warmth and romance, whilst the curvy stairs and bronze aigrette sculpture bring to the lobby a sense of art.

The plan of the Hotel varies in the stretching shape and clean broken line, with a vivid and natural change of step levels from top to bottom echoing to the mountain form. The exterior of the Hotel is a perfect combination of Continental European style in contemporary sense. Delicate details are applied to cornices, walls and balconies.

International Conference Center of Fontainebleau Hotel is located on a hillock of southwest of Fontainebleau Hotel, Nanguo Peach Garden, Foshan City, total area around 20,000 m^2, the project includes Conference Center, Park, Tennis Court etc.. The Center Building is composed of two wings, Conference on the east and Guest Room on the west, there is a center Garden connect to these two parts, and visitors can walk through a corridor from the east side of the conference center to hotel directly. To saving investment and utilizing the height of terrain, the conference center flanked by the hillock, and built a semi-basement here as the logistics and equipment room. There are some independent entrances in the Center and Guest Room, it makes the flow will not be disturbed.

Regarding the layout of plan, there are a 100 seats conference hall, two 50 seats conference hall, a multifunctional entertainment hall and eight small Conference Room in the first floor of Conference Center; a large banquet hall in the second floor. In the Guest Room Building, the first floor is lobby, the second to sixth floor are rooms, vegetation on the roof can keep thermal and heat insulation of the building.

Regarding the facade design, the shape of building is square, the entrance of west side which nearby the intersection present is a semicircular shape layout. Viewers can see the landscape of mountain through the glass wall. The east side is solid surface, the interior design of it focus on the scale and details. Granite cladding-based sculpture and detail treatment, reflecting a kind image and considerate personality of the building.

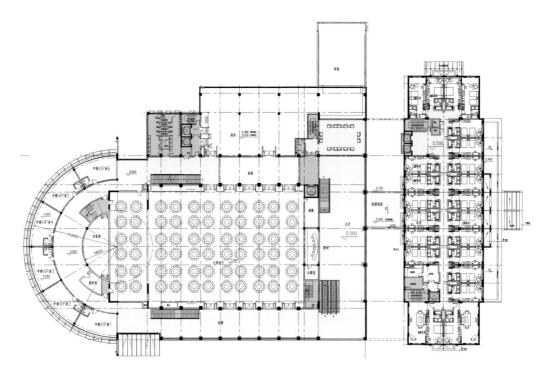

▲ 国际会议中心二层平面图 *Level Two Plan*

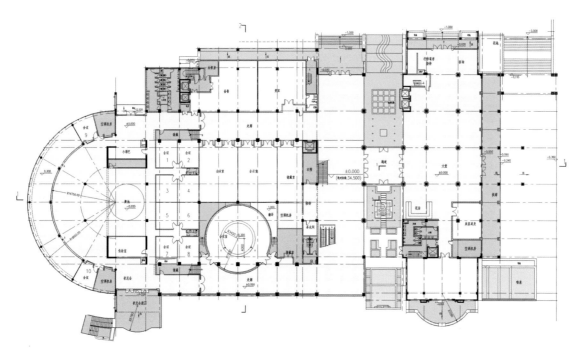

▲ 国际会议中心首层平面图 *Ground Level Plan*

鼎湖·总统御山莊
Ding Hu·Presidential Royal Villas

项目地点： Venue:	肇庆市广利片区 Guangli District, Zhaoqing
设计日期： Design Date:	2011 年 2011
竣工日期： Completion Date:	设计中 Design Stage
用地面积： Project Area:	118,414 平方米（177 亩） 118,414m² (177a)
项目建筑面积： Construction Area:	158,547 平方米 158,547m²
合作设计： Partner:	林峻任、杨智敏、李韬 Lin Junren, Yangzhiming, Litao

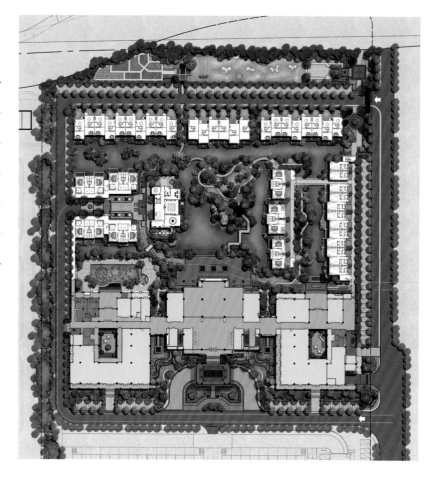

在总统御山庄的设计中，我们以生态酒店文化为创作主题，以自然环境的保护和利用为构思出发点，创作出具有岭南文化特点的高档酒店环境。

酒店采用园林式布局，利用地基高差，北面引鼎湖山水景观进入小区，形成很有特色的山水园林景观。建筑布局采用借景手法，最大限度地将鼎湖山景致引入室内，力求人与自然环境的和谐统一是本项目设计的主要构思。

采用"景观优先"的设计理念，我们将酒店大堂设于二层，居高临下，可近观花园、远眺鼎湖全景，视野十分开阔。沿内庭一侧设生态景观廊，作为酒店客人交通主干道，步行其中，观山赏水，尽赏天人合一之写意的酒店环境。

酒店外形采用现代的材料和工艺，借鉴了中国古建筑构图手法和肇庆当地建筑的特点，整体立面采用三段式布置，基础五层为石材基座，彰显稳固尊贵，中段采用具有中式图案特色的分段式立面设计，结合仿木框玻璃幕墙等设计元素，中段简洁大方，具有住宅公建化特点。顶部借鉴了故宫屋顶建筑造型，采用"重檐歇山"这种最高级的形式，使建筑更显得尊贵大方，有皇家建筑特点。

The design of Presidential Royal Villas adopts ecological hotel culture as the creative theme, conceived from protection and utilization of natural environment, resulting in high-end hotel environment of Lingnan Cultural style.

The garden layout takes advantage of the height difference of the base and introduces the mountain and waterscape of Dinghu to the north into the area, forming the characteristic and unparalleled garden landscape. The approach of "view borrowing" is adopted to furthest induct the scenery of Dinghu, so as to achieve the main conception of the project—the harmony and unity of human and natural environment.

The design evolves around the philosophy "Landscape first". The lobby of the hotel is set on the second floor, the advantageous position ensures an open view of the garden in vicinity and Dinghu in the distance. Ecological landscape corridor lies on one side of the inner courtyard as the main road for hotel guests. Walking through the corridor, the guests could delightedly enjoy the beautiful scenery of the landscape and the perfect "harmony between nature and human" of the hotel environment.

The modern material and technics of the shape take reference from the pattern of Chinese ancient architecture and Zhaoqing local

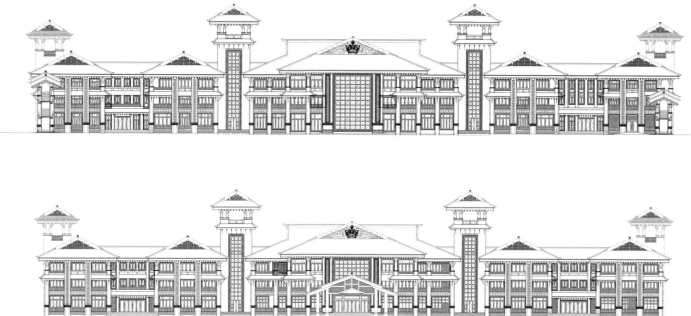

▲ 立面图 Elevation

architecture style. The elevation is designed in three segments: the basic five layers are stone foundation to ensure steadiness and nobleness; the middle part uses the sectional elevation design with Chinese feature patterns combining with design elements like imitated-wood framed glass curtain wall, which is simple and generous featuring public construction of residence; the top applies the architecture form of the rooftops in the Forbidden City—the top-drawer form of gable and hip roof with multiple eaves, endowing the hotel with dignity, decency and touch of royal architecture.

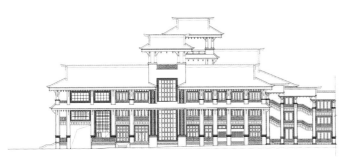

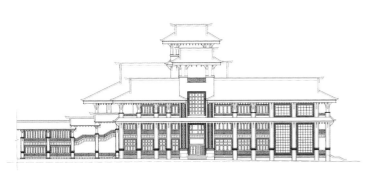

▲ 立面图 *Elevation*

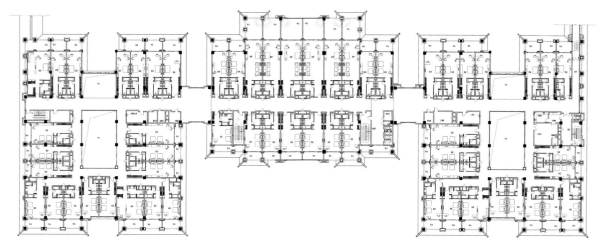

▲酒店三层平面图 *Level Three Plan*

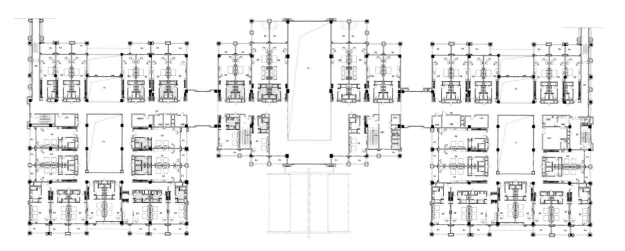

▲酒店二层平面图 *Level Two Plan*

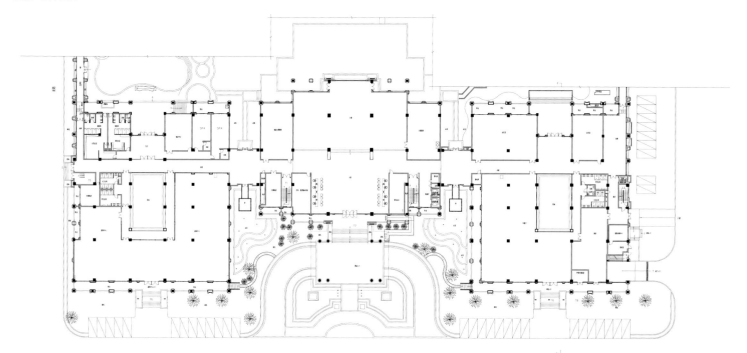

▲酒店首层平面图 *Ground Level Plan*

中山大学东校区图书馆
Library of the East Campus, Sun Yat-sen University

中山大学东校区公共教学楼
Public Classroom Building of East Campus, Sun Yat-sen University

中山大学东校区基础实验楼
Fundamental Laboratory Building of East Campus, Sun Yat-sen University

中山大学东校区工学院
School of Polytechnics of East Campus, Sun Yat-sen University

广州科技职业学院
Guangzhou Vocational College of Science and Technology

广东财经大学华商学院
Huashang College of Guangdong University of Finance & Economics

教育建筑 EDUCATIONAL BUILDINGS

中山大学东校区法学院
School of Law of East Campus, Sun Yat-sen University

中山大学东校区行政学院
School of Administration of East Campus, Sun Yat-sen University

中山大学东校区传播与设计学院
School of Communications and Design of East Campus, Sun Yat-sen University

中山大学东校区行政会议中心
Administration & Conference Center of East Campus, Sun Yat-sen University

中山大学东校区微纳尺度材料及生命科学实验大楼
Nanoscale Materials & Life Sciences Laboratory Building of East Campus, Sun Yat-sen University

中山大学东校区工科实验楼与药学院楼
Engineering Laboratory Building & Pharmaceutical Building of Sun Yat-sen University, Guangzhou Higher Education Mega Center

南昌市安义县教育园
Education Park of Anyi County, Nanchang

桂林航天工业学院
Guilin University of Aerospace Technology

中山大学东校区图书馆
Library of the East Campus, Sun Yat-sen University

项目地点： Venue:	广州市小谷围岛 Xiaoguwei Island, Guangzhou
设计日期： Design Date:	2003 年 2003
竣工日期： Completion Date:	2004 年 2004
项目建筑面积： Construction Area:	35,000 平方米 35,000m²
曾获奖项： Awards:	①广东省优秀建筑设计一等奖 ②住房和城乡建设部优秀建筑设计三等奖 ③住房和城乡建设部建筑学会创作奖·佳作奖 ① First Prize, Excellent Architectural Design of Guangdong Province ② Third Prize, Excellent Design of Ministry of Construction ③ Honorable Mention Award, Architectural Society of China Architecture Creative Awards, by Ministry of Construction
合作设计： Partner:	郭明卓、郑启皓 Guo Mingzhuo, Zheng Qihao

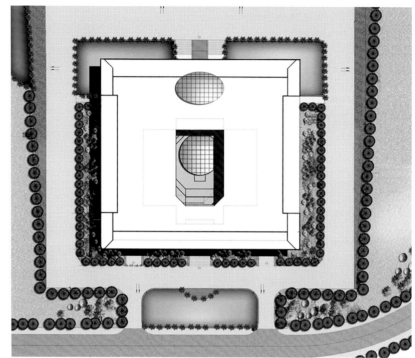

▲总平面图 *Master Plan*

中山大学东校区图书馆位于中山大学东校区主广场最南端，是中山大学标志性建筑物。设计注重现代图书馆应具有的全方位、高效率与高质量的特点，做到各功能分区明确，联系便捷，人流和书流完全分开。

共享中庭贯穿了全部阅读空间，而主交通空间及读者休息空间都与共享绿化中庭相伴。共享的绿化露台为读者提供了立体的阅读空间，以及交流氛围。这样既有静谧与而沉稳的主阅读空间，又有生气勃勃的交流空间，是在现代化大学图书馆设计中的体现，是更人性化、感情化的空间形式。

图书馆设计中运用了较大的造型尺度，如大门框、翻开的书页形象以及入口高柱式和大台阶、跌水池等，都是追求设计的简约舒展，从而符合中山大学作为华南第一学府之恢宏大气的气势，也是对于中山大学中大主入口中轴线的一种呼应。

从远处看，建筑的外部空间形态犹如一个架立于天地间抽象的辕门。校园中轴线正对着辕门而过，两幅左右对称的巨型弧形墙面犹如两扇大门，敞开着，形成很好的导向，并且围合形成一个抬升的图书馆的入口小广场，弧形墙面更像张开的双臂欢迎莘莘学子进入这智慧学术之门。图书馆北面主入口门厅设于二层，层层向上的台阶暗喻"书山有路勤为径"，而中轴线穿越图书馆后，遥指大学城中心公园，暗示了"学无止境"的至高境界。

The symbolic architecture of the Library sits at the south end of the East Campus of Sun Yat-sen University, the design focus on the features possessed by a modern library such as omni-functional, high-efficient and high-quality. Each functional area is set distinct and conveniently connected to each other and the people stream and bookshelves are completed separated.

The public atrium runs through every reading area, enabling the main traffic space to share the greenery with reader's relaxation space. The shared green terrace provides three-dimensional space and communication environment for readers. Such tranquil and staid reading area, integrated with vivid interaction space, marks the sharing and intercommunion way of a modern library, which is also a more humane and sensational form.

Bigger scale in shape is utilized in the design to pursue simplicity and extension, such as large doorframe, image of an open book, columns at the entrance, big steps and plunge pool. It matches the grandness and generosity as "the best university of South China", and also echoes the central axis of the main entrance of Sun Yat-sen University.

Seen from a distance, the exterior of the architecture is like the outer gate of government official erecting between the sky and the earth. The central axis of the campus runs through the doorframe, and led by a pair of symmetrical giant curvy doors, resembles to open arms

welcoming the students into the gate of wisdom, and forms the raised entrance square of the library. The entry hall in the north is set on the second floor. The steps leading upstairs indicates the motto "Diligence is the key leading to the sea of knowledge". The central axis points to the Central Park of Higher Education Mega Center after traversing the library, implying the supreme realm of "Live and learn".

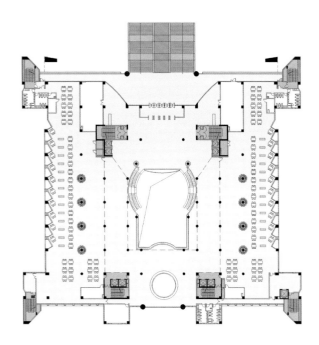

▲ 二层平面图 *Level Two Plan*

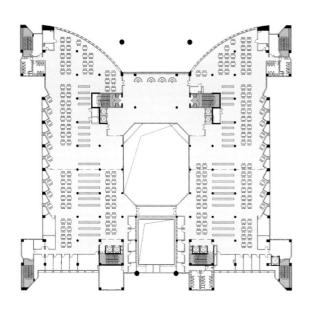

▲ 四层平面图 *Level Four Plan*

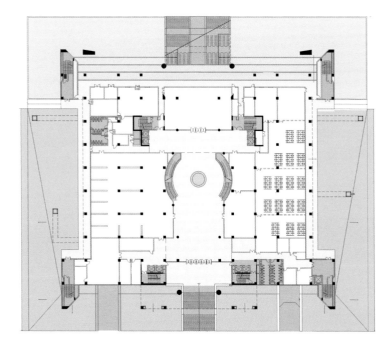

▲ 首层平面图 *Ground Level Plan*

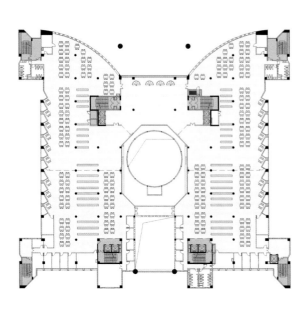

▲ 三层平面图 *Level Three Plan*

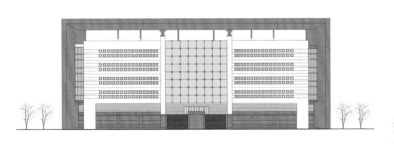

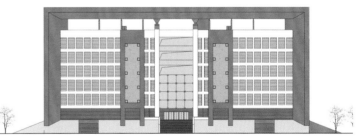

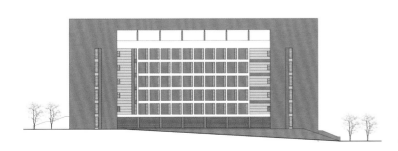

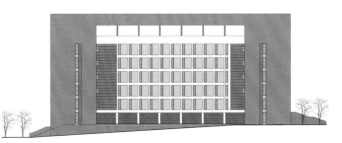

▲ 上图： 北立面图 *Top: North Elevation*
▲ 下图： 西立面图 *Bottom: West Elevation*

▲ 上图： 南立面图 *Top: South Elevation*
▲ 下图： 东立面图 *Bottom: East Elevation*

中山大学东校区公共教学楼

Public Classroom Building of East Campus, Sun Yat-sen University

项目地点： Venue:	广州市小谷围岛 Xiaoguwei Island, Guangzhou
设计日期： Design Date:	2003 年 2003
竣工日期： Completion Date:	2004 年 2004
项目建筑面积： Construction Area:	37,715 平方米 37,715m²
合作设计： Partner:	① 广东省优秀建筑设计一等奖 ② 住房和城乡建设部优秀建筑设计三等奖 ③ 住房和城乡建设部建筑学会创作奖·佳作奖 ① First Prize, Excellent Architectural Design of Guangdong Province ② Third Prize, Excellent Design of Ministry of Construction ③ Honorable Mention Award, Architectural Society of China Architecture Creative Awards, by Ministry of Construction
合作设计： Partner:	郭明卓、蔡展刚 Guo Mingzhuo, Cai Zhangang

中山大学东校区公共教学楼位于校园主广场东侧，由五座不同类型的教室大楼和围绕着的庭园组成。各大楼主入口是利用阶梯教室架空层形成的主入口，结合有雕塑感的白色衬底的红砖山墙造型，形成具有韵律感的立面阵列。

Public Classroom Building of East Campus is situated at the east end of the Main Square, composed by surrounding courtyard and five classroom buildings of various types. The main entrance of the buildings is formed by the stilt floor of amphitheatres, in combination with sculptural gable wall made of red bricks in white background, bringing forth the rhythmic elevation array.

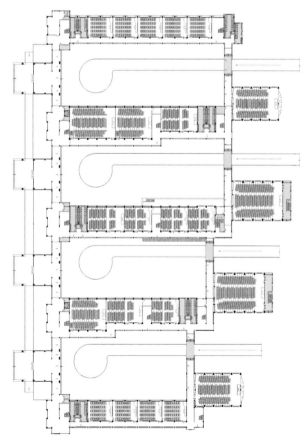

▲ 首层平面图 *Ground Level Plan*

中山大学东校区法学院

School of Low of East Campus, Sun Yat-sen University

项目地点： Venue:	广州市小谷围岛 Xiaoguwei Island, Guangzhou
设计日期： Design Date:	2003 年 2003
竣工日期： Completion Date:	2004 年 2004
项目建筑面积： Construction Area:	19,688 平方米 19,688m²
曾获奖项： Awards:	① 广东省优秀建筑设计一等奖 ② 住房和城乡建设部优秀建筑设计三等奖 ③ 住房和城乡建设部建筑学会创作奖·佳作奖 ① First Prize, Excellent Architectural Design of Guangdong Province ② Third Prize, Excellent Design of Ministry of Construction ③ Honorable Mention Award, Architectural Society of China Architecture Creative Awards, by Ministry of Construction
合作设计： Partner :	郭明卓、吕向红 Guo Mingzhuo, Lü Xianghong

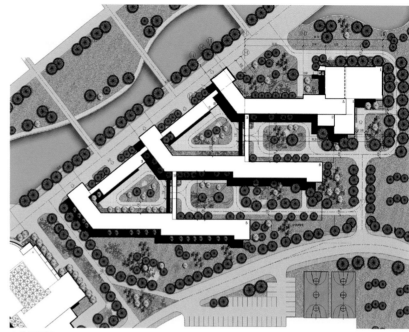

▲ 总平面图 *Master Plan*

中山大学东校区法学院位于校园生态轴南侧，用地依山靠水，环境优美。行政学院由三组外廊式建筑组成，各学院的主入口设在生态景观轴上临水的一侧，各塔式门楼通过外廊的相连形式，有机地组合成韵律感极强的整体。

School of Low of East Campus stands on the south side of ecological axis of Sun Yat-sen University. It's flanked by mountain and water, looks liking a scenic spot. Three Veranda style architectures consisted of the main building, and the main entrance sits by the waterscape of the ecological landscape axis. The tower-shape gateways link to each other by verandas and organically assemble into highly rhythmic entity.

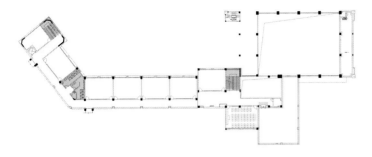

▲ 二层平面图 Level Two Plan

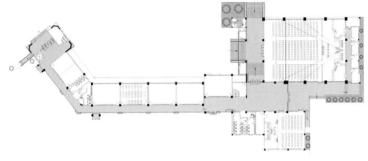
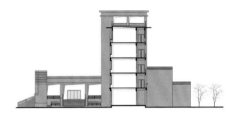

▲ 首层平面图 Ground Level Plan　　　　　▲ 立面图 Elevation

中山大学东校区行政学院
School of Administration of East Campus, Sun Yat-sen University

项目地点： Venue:	广州市小谷围岛 Xiaoguwei Island, Guangzhou
设计日期： Design Date:	2003 年 2003
竣工日期： Completion Date:	2004 年 2004
项目建筑面积： Construction Area:	18,000 平方米 18,000m²
曾获奖项： Awards:	① 广东省优秀建筑设计一等奖 ② 住房和城乡建设部优秀建筑设计三等奖 ③ 住房和城乡建设部建筑学会创作奖·佳作奖 ① First Prize, Excellent Architectural Design of Guangdong Province ② Third Prize, Excellent Design of Ministry of Construction ③ Honorable Mention Award, Architectural Society of China Architecture Creative Awards, by Ministry of Construction
合作设计： Partner:	郭明卓、李明 Guo Mingzhuo, Li Ming

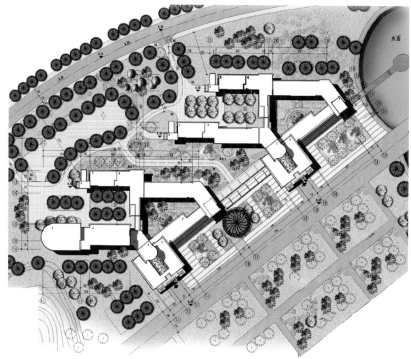

▲ 总平面图 *Master Plan*

中山大学东校区行政学院临江而建，位于校园生态轴的北侧，平面采用院落式组合布局，保留了两棵百年老树。入口为门框式造型，结合新型材料的运用，建筑造型具有雕塑美感。

School of Administration of East Campus was built by the river on the north side of the ecological axis of Sun Yat-sen University. The plan adopts courtyard arrangement and retains two-century-old trees. The doorframe-shape entrance, to use new materials, the architecture form was designed full of sculptural beauty.

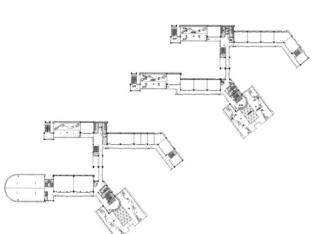

▲ 四层平面图　*Level Four Plan*

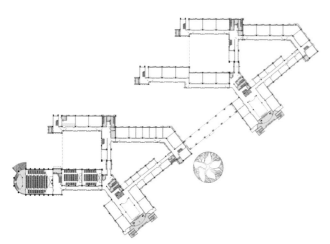

▲ 首层平面图　*Ground Level Plan*　　　　　▲ 立面图　*Elevation*

中山大学东校区传播与设计学院

School of Communications and Design of East Campus, Sun Yat-sen University

项目地点： **Venue:**	广州市小谷围岛 Xiaoguwei Island, Guangzhou
设计日期： **Design Date:**	2003 年 2003
竣工日期： **Completion Date:**	2004 年 2004
项目建筑面积： **Construction Area:**	6,185 平方米 6,185m²
曾获奖项： **Awards:**	① 广东省优秀建筑设计一等奖 ② 住房和城乡建设部优秀建筑设计三等奖 ③ 住房和城乡建设部建筑学会创作奖·佳作奖 ① First Prize, Excellent Architectural Design of Guangdong Province ② Third Prize, Excellent Design of Ministry of Construction ③ Honorable Mention Award, Architectural Society of China Architecture Creative Awards, by Ministry of Construction
合作设计： **Partner :**	郭明卓、吕向红 Guo Mingzhuo, Lü Xianghong

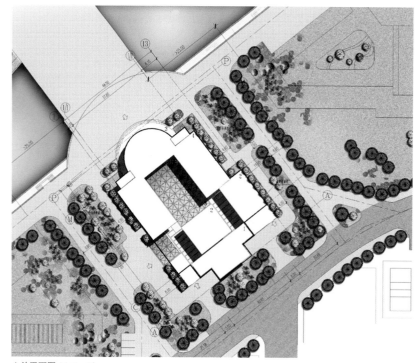

▲总平面图 *Master Plan*

中山大学东校区传播与设计学院由小剧院和教学楼连体构成，位于校园生态轴中央，与对面保留山体相互对应，整体以方、圆为构图母题，造型自由活泼，具有时代气息，以体现传播学院信息化、时尚化的特点。

School of Communications and Design of East Campus lies at the center of ecological axis of Sun Yat-sen University constituted of little theatre and classroom building. The school corresponds to the opposite retained mountain, and employs square and circle as the theme plan, with the frees-style and cotemporary shape, representing the extraordinary trait of informationization and fashion in respect of Communications and Design.

▲ 剖面图 Section ▲ 立面图 Elevation

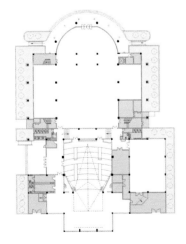 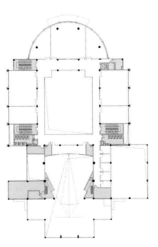  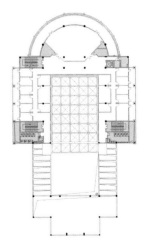

▲ 首层平面图 Ground Level Plan   ▲ 二层平面图 Level Two Plan   ▲ 三层平面图 Level Three Plan   ▲ 四层平面图 Level Four Plan

中山大学东校区基础实验楼
Fundamental Laboratory Building of East Campus, Sun Yat-sen University

项目地点： **Venue:**	广州市小谷围岛 Xiaoguwei Island, Guangzhou
设计日期： **Design Date:**	2003 年 2003
竣工日期： **Completion Date:**	2004 年 2004
项目建筑面积： **Construction Area:**	25,680 平方米 25,680m²
曾获奖项： **Awards:**	① 广东省优秀建筑设计一等奖 ② 住房和城乡建设部优秀建筑设计三等奖 ③ 住房和城乡建设部建筑学会创作奖·佳作奖 ① First Prize, Excellent Architectural Design of Guangdong Province ② Third Prize, Excellent Design of Ministry of Construction ③ Honorable Mention Award, Architectural Society of China Architecture Creative Awards, by Ministry of Construction
合作设计： **Partner:**	郭明卓、周茂 Guo Mingzhuo, Zhou Mao

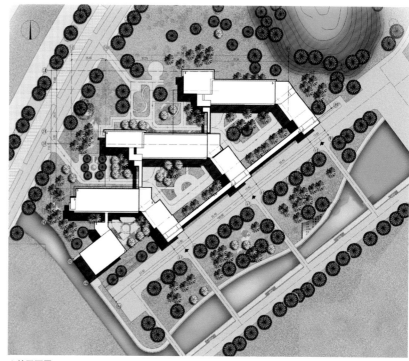

▲总平面图 *Master Plan*

中山大学东校区基础实验楼设计时充分考虑本项目与东侧小山丘及相邻工学院楼群的协调关系，同时也注重利用地处两侧的江面及中心绿地景观，实验楼采用园林式布局，创造良好的教学环境，临中心绿地一侧利用实验楼山墙造型，塑造具有韵律感的建筑形象。

The design of Fundamental Laboratory Building has thoroughly considered the coordination between the project and the eastern hill together with the adjacent buildings of school of polytechnics, and properly utilized river-scape and central greening on both sides. Garden layout is introduced to create delightful learning environment. Gable wall is designed for the building on the side of the greenbelt to build up the cadenced architecture image.

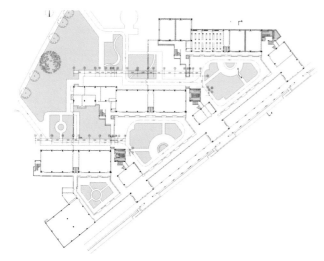

▲首层平面图 *Ground Level Plan*

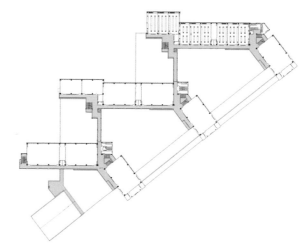

▲二～五层平面图 *Level Two ~ Five Plan*

中山大学东校区工学院
School of Polytechnics of East Campus, Sun Yat-sen University

项目地点： Venue:	广州市小谷围岛 Xiaoguwei Island, Guangzhou
设计日期： Design Date:	2003 年 2003
竣工日期： Completion Date:	2004 年 2004
项目建筑面积： Construction Area:	22,000 平方米 22,000m²
曾获奖项： Awards:	① 广东省优秀建筑设计一等奖 ② 住房和城乡建设部优秀建筑设计三等奖 ③ 住房和城乡建设部建筑学会创作奖·佳作奖 ① First Prize, Excellent Architectural Design of Guangdong Province ② Third Prize, Excellent Design of Ministry of Construction ③ Honorable Mention Award, Architectural Society of China Architecture Creative Awards, by Ministry of Construction
合作设计： Partner:	郭明卓、钟献荣 Guo Mingzhuo, Zhong Xianrong

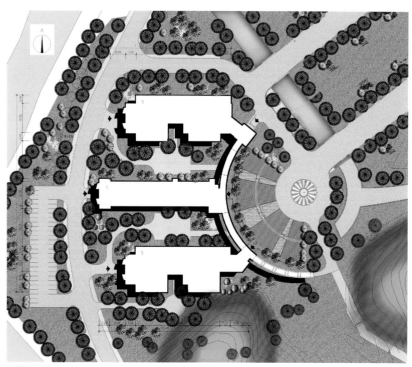

▲总平面图 *Master Plan*

中山大学东校区工学院处于中山大学校园生态轴尽端，在平面上利用弧形廊道将三栋建筑物构成有机联系的建筑群，并在室外围合成半圆形的室外剧场，立面上利用标志塔延续了校本部钟塔的文化特征，在校区生态轴空间上起对景作用，在建筑群的外形上起制高点作用。室外大台阶直上二层大堂，可观赏生态轴全景。

School of Polytechnics of East Campus stands at the end of the ecological axis of Sun Yat-sen University. On plane, curvy corridors tie three buildings together to form the complex and enclose a semicircular outdoor theatre. On elevation, the symbolic tower maintains the cultural feature of clock tower of Sun Yat-sen University, functions as opposite scenery on the ecological axis and commanding point in the shape of architectures. The big outdoor steps lead to the lobby on the second floor, from where the entire ecological axis can be viewed.

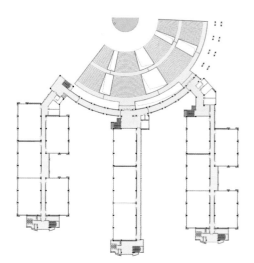

▲二~五层平面图 *Level Two ~ Five Plan*

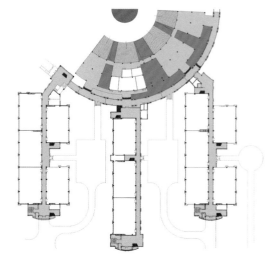

▲首层平面图 *Ground Level Plan*

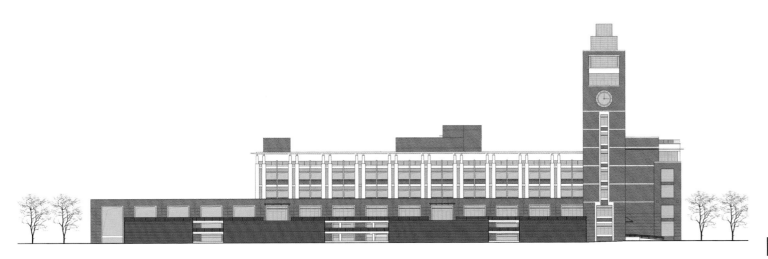

▲ 立面图 *Elevation*

中山大学东校区微纳尺度材料及生命科学实验大楼

Nanoscale Materials & Life Sciences Laboratory Building of East Campus, Sun Yat-sen University

项目地点： Venue:	广州市小谷围岛 Xiaoguwei Island, Guangzhou
设计日期： Design Date:	2003 年 2003
竣工日期： Completion Date:	2005 年 2005
项目建筑面积： Construction Area:	46,518 平方米 46,518m²
曾获奖项： Awards:	① 广东省优秀建筑设计二等奖 ② 住房和城乡建设部优秀建筑设计三等奖 ① Second Prize, Excellent Architectural Design of Guangdong Province ② Third Prize, Excellent Design of Ministry of Construction
合作设计： Partner:	郭明卓、黄考颖、黄惠青 Guo Mingzhuo, Huang Kaoying, Huang Huiqing

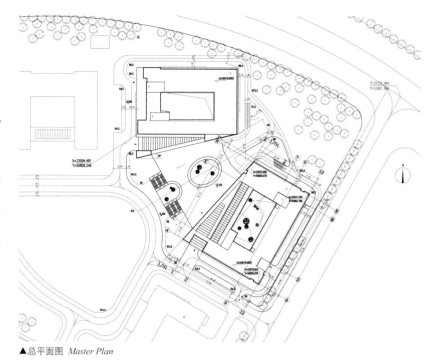

▲ 总平面图 *Master Plan*

▲ 黄劲手绘效果图 *Hand Drawing, Huang Jin*

中山大学东校区微纳尺度材料及生命科学实验大楼为国家级科研楼，包括国家重点实验室、工作室、仪器室等功能用房。建筑采用对称庭园式布局，中间设大型公共平台相连，使两组建筑有机组合，既满足使用功能要求，又使建筑群组外观协调统一。

本工程主入口位于二层公共平台，平台下为停车场，后勤入口位于首层，交通方便。立面采用简洁明快的现代手法，靠平台的立面以有想象力的弧面作为入口标志，营造出传承文明、开拓创新的文化氛围和现代气息。

Nanoscale Materials & Life Sciences Laboratory Building of East Campus is the state-level scientific research building, including functional spaces of state key laboratory, workshop and apparatus rooms etc. The architecture is designed in symmetrical courtyard layout, combining two buildings via a large public platform in the middle, hence the functional requirement is met, and the peripheral looking of the architecture could be harmonious and united.

The traffic becomes very convenient as the main entrance is set on the public platform on the second floor, underneath is the car park, and the logistics entrance is on the ground floor. Vivid and simple, the contemporary approach of the elevation uses imaginative cambered surfaces as entry symbol, presenting the abundance heritage, innovative culture and contemporary sense of Sun Yat-sen University.

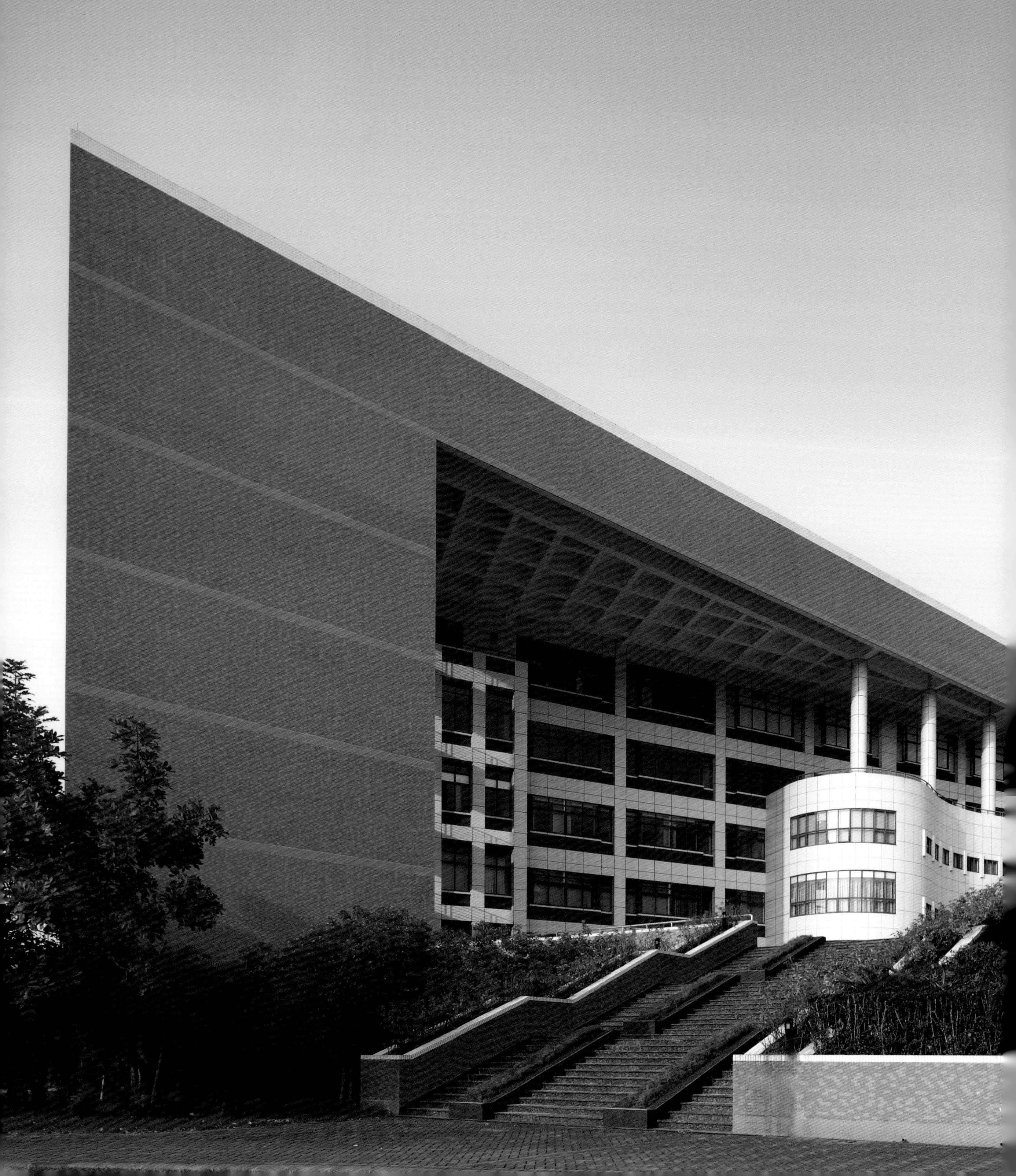

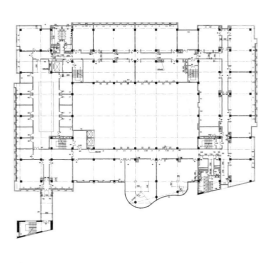

▲三层平面图 *Level Three Plan*

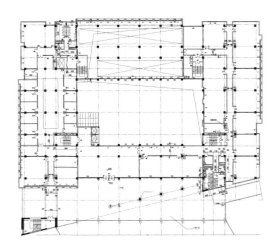

▲二层平面图 *Level Two Plan*

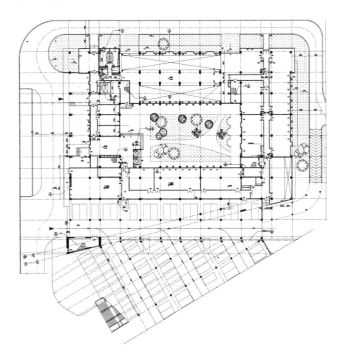

▲首层平面图 *Ground Level Plan*

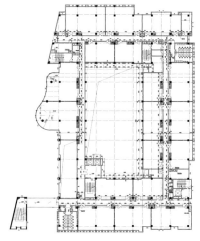
▲ 三层平面图 Level Three Plan

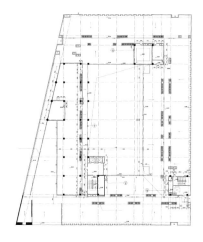
▲ 屋面平面图 Roof Plan

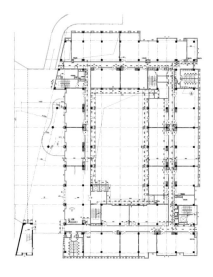
▲ 二层平面图 Level Two Plan

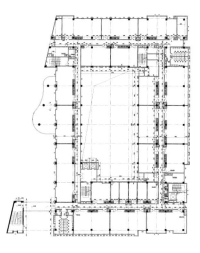
▲ 五层平面图 Level Five Plan

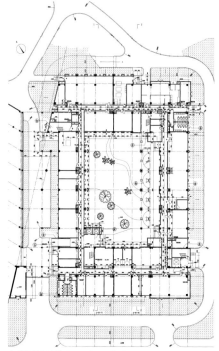
▲ 首层平面图 Ground Level Plan

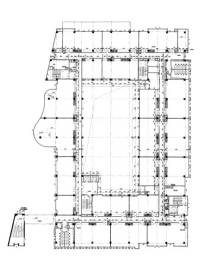
▲ 四层平面图 Level Four Plan

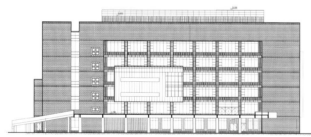

▲ 西立面图 *West Elevation*

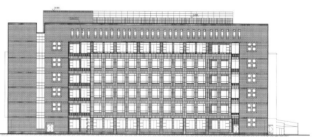

▲ 东立面图 *East Elevation*

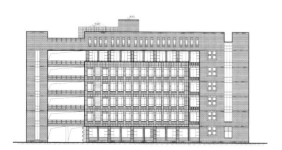

▲ 南立面图 *South Elevation*

▲ 北立面图 *North Elevation*

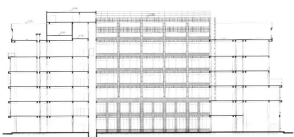

▲ 1-1 剖面图 1-1 *Section*

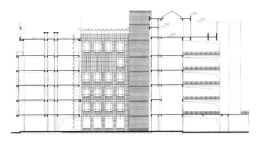

▲ 2-2 剖面图 2-2 *Section*

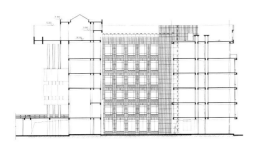

▲ 3-3 剖面图 3-3 *Section*

▲ 4-4 剖面图 4-4 *Section*

中山大学东校区行政会议中心

Administration & Conference Center of East Campus, Sun Yat-sen University

项目地点： Venue:	广州市小谷围岛 Xiaoguwei Island, Guangzhou
设计日期： Design Date:	2005 年 2005
竣工日期： Completion Date:	2009 年 2009
建筑面积： Project Area:	12,930 平方米 12,930m²
曾获奖项： Awards:	① 广东省优秀建筑设计一等奖 ② 住房和城乡建设部优秀建筑设计三等奖 ① First Prize, Excellent Architectural Design of Guangdong Province ② Third Prize, Excellent Design of Ministry of Construction
合作设计： Partner:	郭明卓、郑晓山、关小梅 Guo Mingzhuo, Zheng Xiaoshan, Guan Xiaomei

　　中山大学东校区行政会议中心由于功能有别，根据日照、通风、防火、卫生和交通等相应要求的差异，建筑总平面布置采用化整为零的处理手法，将行政办公楼和会议中心两大功能部分分开处理，避免了动静区域的相互干扰；中间再用一连廊和架空屋盖将两部分连为一体，增强建筑的联系和整体感。

　　北面的行政楼以一个传统的"回"字形的平面为基础，方正实用，利于塑造较为庄重的办公形象；南面的会议中心则呈较现代的扇形倒锥体，大小会议厅室有机组合，空间活泼而富有动感，并配合场地的水体环境形成灵动的亲水建筑空间。传统和现代的元素在此有了初次的碰撞，悬空连廊和共享内庭的设置，则成为两者之间顺利的过渡。

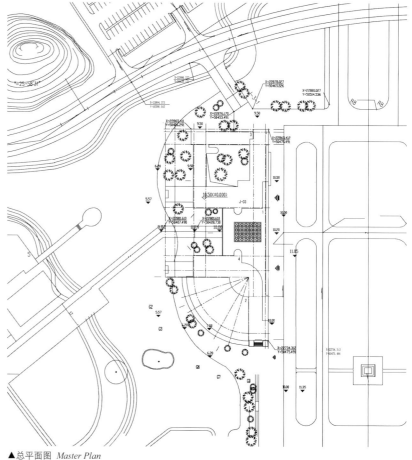

▲总平面图 *Master Plan*

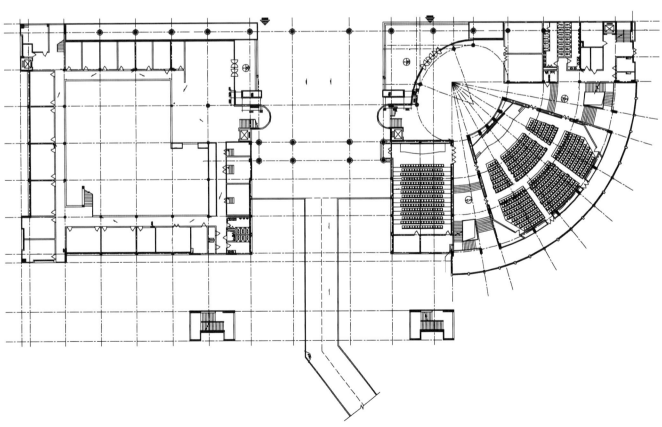

▲ 首层平面图 *Ground Level Plan*

222

According to the various requirements on daylight, ventilation, fireproofing, sanitation and traffic, "break up the whole into parts" is manipulated for the master plan layout of Administration & Conference Center of East Campus due to its special functional needs. The Administration Office Building and the Conference Center are separated to avoid interference from activity area to professional area. The two parts are tied by a sky bridge and overhead roof in-between to enhance the connection and integrity of the architecture.

The north Administration Building is based on a " 回 " structure of the traditional character, which is practical and suitable for creating a solemn image of office environment. The Conference Center is in the shape of modern sectorial cone placed upside down. Combination of large and small conference rooms engender vivid and dynamic space, which creates delicate Water-affinity architecture complying with water environment of the venue. The crash of traditional and contemporary elements is premiered, while the overhead sky bridge and public inner courtyard construct a smooth transition between both.

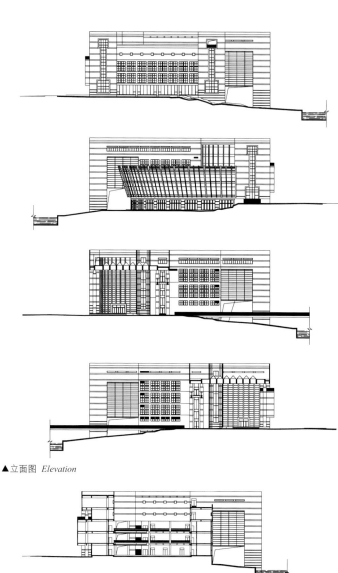

▲ 立面图 *Elevation*

▲ 立面、剖面图 *Elevation & Section*

▲ 剖面图 *Section*

中山大学东校区工科实验楼与药学院楼
Engineering Laboratory Building & Pharmaceutical Building of East Campus, Sun Yat-sen University

项目地点： Venue:	广州市小谷围岛 Xiaoguwei Island, Guangzhou
设计日期： Design Date:	2005 年 2005
竣工日期： Completion Date:	2009 年 2009
用地面积： Project Area:	34,871 平方米 34,871m²
项目建筑面积： Construction Area:	42,700 平方米 42,700m²
曾获奖项： Awards:	广东省优秀建筑设计二等奖 Second Prize, Excellent Architectural Design of Guangdong Province
合作设计： Partner:	郭明卓、郑晓山、关小梅 Guo Mingzhuo, Zheng Xiaoshan, Guan Xiaomei

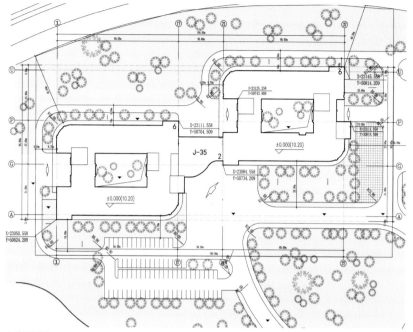

▲总平面图 *Master Plan*

本工程项目位于中山大学东校区，西端为工程学科实验楼，东端为药学院实验楼，两栋实验楼均采用围合式内庭院设计，平面布局合理，满足功能使用要求。建筑首层为行政办公及部分实验室，二层以上主要为各种实验室、学术交流室、资料室及标准功能用房。

中山大学是一座具有悠久历史和文化的高等学府，校区的大部分建筑伴随中山大学的成长，积累了深厚的文化底蕴。本工程建筑造型设计中，充分考虑了这一因素，传承文明，开拓创新，从原有建筑中提炼出中山大学的文化精髓，运用现代的处理手法，令建筑稳重大方，现代又具有较高的文化品位。采用简洁明快的立面设计手法，砖红色与白色的墙面相互呼应，营造出校园的文化氛围和现代的气息。

This project is located within the East Campus of Sun Yat-sen University, with the Engineering Laboratory Building lies at its west end, and Pharmaceutical Building at its east end. Both Laboratory Buildings are designed in the form of enclosed inner courtyard, and the reasonable layout could fulfill all functional requirements. The Administration Office and part of the laboratories are set on the first floor, and the laboratories, Academic Exchanges Rooms, Reference Rooms and Functional Rooms are accommodated on the second floor.

Being an institution of higher education of long history and renowned culture, most architecture come through along with Sun Yat-sen University and share the solid cultural background. This factor is also considered in the shape design of the architecture for culture inheritance and innovation, so as to extract the culture essence from the original building. Modern manipulation is applied to make the architecture look calm, generous, contemporary yet in a taste of high education. The simple and vibrant elevation design is presented with the echo of brick red and white walls, which properly creates the special cultural atmosphere and contemporary touch for Sun Yat-sen University.

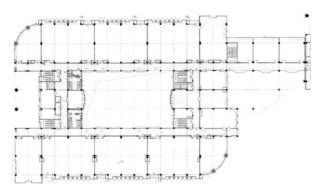

▲ A 栋二层平面图 *Level Two Plan of A*

▲ B 栋二层平面图 *Level Two Plan of B*

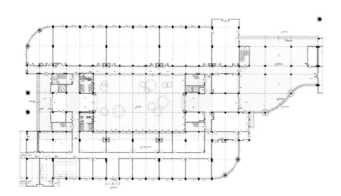

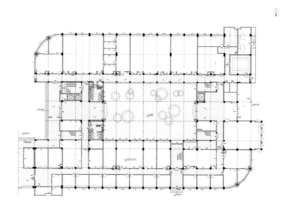

▲ A 栋首层平面图 *Ground Level Plan of A*

▲ B 栋首层平面图 *Ground Level Plan of B*

广州科技职业技术学院
Guangzhou Vocational Collage of Science and Technology

项目地点： Venue:	广州从化市钟落潭镇 Zhongluotan Town, Conghua, Guangzhou
设计日期： Design Date:	2008 年 2008
竣工日期： Completion Date:	2010 年 2010
项目建筑面积： Construction Area:	800,096 平方米 800,096m²
用地面积： Project Area:	619,005 平方米 619,005m²
合作设计： Partner:	黄文成、冯险峰、颜立华、肖永强 Eng Bon Seng, Feng Xianfeng, Yan Lihua, Xiao Yongqiang

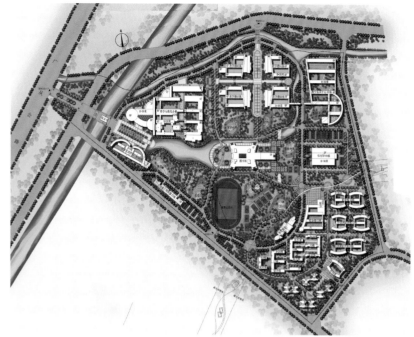

▲规划设计总平面图 *Master Planning + Urban Design*

广州科技职业技术学院规划设计以岭南生态的校园文化为创作主题，以自然环境的保护和利用为构思出发点，创造出具有浓厚岭南特色的校园环境。

校园建筑围绕校园生态文化轴布置，生态文化轴贯穿校园建筑各组团，将教学区、生活区、体育运动区连为一体，成为建筑室内外空间的联系纽带。生态轴上，一系列的水景、绿化、广场，为师生提供交往、休闲的活动空间。

图书馆外形极具雕塑美感，以突现其标志性建筑的意义。设计中较大的造型尺度，如大门框，张开的书页等，追求简约舒展，是对校园中轴线的呼应。

综合楼采用客家围屋的设计构思，平面由两组环形宿舍楼组成，利用形体的内聚性，为了学生创造更密切交往场所，增强学生凝聚力，让学生有围屋般"家"的感受。

The design of Guangzhou Vocational College of Science and Technology evolves around the campus culture of Lingnan ecology, starting from the idea of protection and utilization of natural environment, to create the campus ambience of strong Lingnan feature.

The buildings are planned surrounding the campus ecological cultural axis, which acts as the linkage of interior and exterior space by running through various functions of architecture and connecting Classroom Area, Living Quarters and Sports Area together. A serial of waterscape, greening and squares provide intercourse and recreation space for teachers and students on the ecological axis.

The form of the library is full of sculptural beauty to outstand the importance of this symbolic building. Bigger scale in shape is utilized in the design to pursue simplicity and extension, such as large doorframe and image as an open book, which also echoes the central axis of the campus.

The design concept of the Complex Building is inspired by the construction of Hakka Enclosed Houses. Two sets of circular dormitory buildings create intimate interaction space for the students by the cohesion of its form, which effectively enhance the team spirit and bestow home-like feeling for the students.

▲黄劲手绘的综合楼效果图 *Hand Drawing of Complex Building, Huang Jin*

▲ 鸟瞰效果图 Aerial View

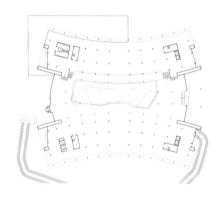

▲图书馆首层平面图　Ground Level Plan of Library

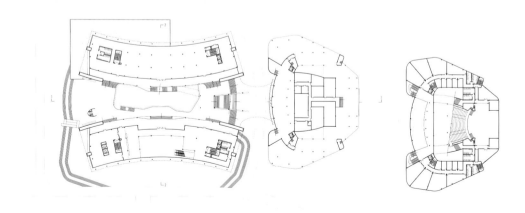

▲图书馆二层平面图　Level Two Plan of Library　　　▲办公楼首层和二层平面图　Ground Level Plan & Level Two Plan of Office Building

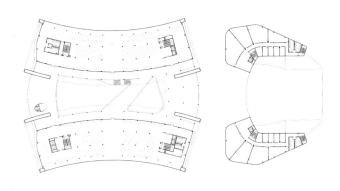

▲图书馆和办公楼五层平面图 *Level Five Plan of Library & Office Building*

▲办公楼屋面平面图 *Roof Plan of Office Building*

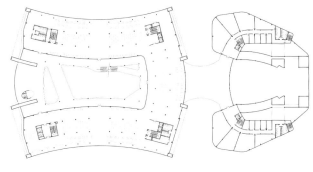

▲图书馆和办公楼四层平面图 *Level Four Plan of Library & Office Building*

▲图书馆屋面层平面图 *Roof Plan of Library*

▲办公楼六层平面图 *Level Six Plan of Office Building*

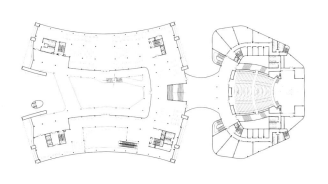

▲图书馆和办公楼三层平面图 *Level Three Plan of Library & Office Building*

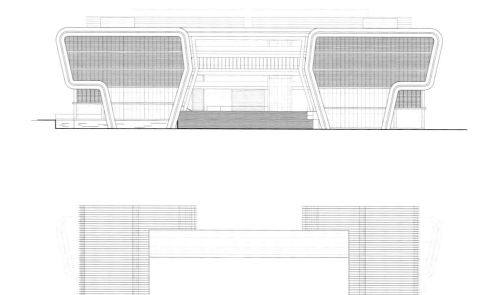

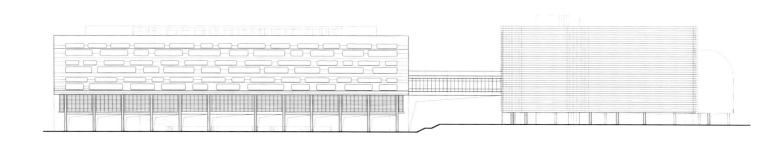

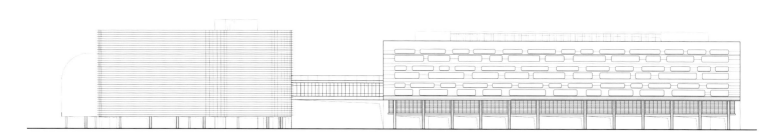

▲ 立面图 Elevation

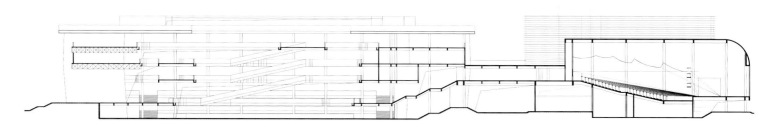

▲ 剖面图 Section

广东财经大学华商学院

Huashang College, Guangdong University of Finance & Economics

项目地点： Venue:	广州增城市荔新大道 Lixin Avenue, Zengcheng, Guangzhou
设计日期： Design Date:	2005年 2005
竣工日期： Completion Date:	2006年 2006
用地面积： Project Area:	518,000平方米 518,000m^2
项目建筑面积： Construction Area:	500,000平方米 500,000m^2
曾获奖项： Awards:	广东省优秀建筑设计三等奖 Third Prize, Excellent Architectural Design of Guangdong Province
合作设计： Partner:	黄军鹏、方良兵、黎就华、黄国明、徐考珍、杨智敏 Huang Junpeng, Fang Liangbing, Li Jiuhua, Huang Guoming, Xu Kaozhen, Yang Zhimin

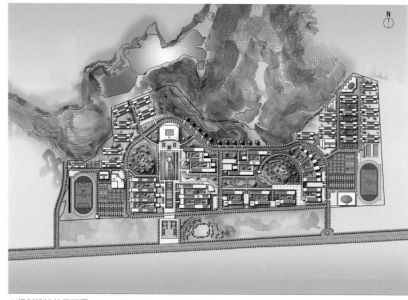

▲规划设计总平面图 *Master Planning + Urban Design*

广东财经大学华商学院建筑围绕校园生态轴布置，轴线正中的图书馆为校园标志性建筑，其两侧的弧形外墙让人联想到一本打开的书本。主广场东侧为基础实验楼，西侧为公共教学楼。建筑造型有强烈的雕塑感、韵律感，砖红色的外墙赋予学院浓厚的文化艺术气息。

The architecture of Huashang College is arranged around the ecological axis on the campus. The library in the center is established as the symbolic building with curvy walls at both sides, associating with an open book. Fundamental Laboratory Building locates at the east side of the main square, and the Public Classroom Building sits in the west. The architecture shape is presented in strong sculptural and rhythmic sense, the brick-red wall endows the campus with sophisticated aura mixed with culture and art.

▲ 鸟瞰效果图 *3D Rendering*

南昌市安义教育园
Education Park of Anyi County, Nanchang

项目地点： Venue:	南昌市安义县 Anyi County, Nanchang
设计日期： Design Date:	2011 年 2011
用地面积： Project Area:	204,001 平方米 204,001m²
项目建筑面积： Construction Area:	约 12 万平方米 Around 120,000m²
曾获奖项： Awards:	设计竞赛中标 Bid Winner of Design Competition
合作设计： Partner :	冯险峰 Feng Xianfeng

汇集了七所学校的安义县教育园，是国家首批建设的县级教育园区。设置教育园区文化生态轴作为规划结构骨架，是方案设计的主要构思，建筑外墙以砖红色为主色，结合轻质钢结构及通透材料的使用，外形现代时尚，具有雕塑美感和文化气息。建筑围绕南北向的校园文化生态轴布置，在生态轴中央布置图书馆、青少年活动中心、体育馆等公共建筑，作为资源共享区为七所学校提供服务。

Consisted of sever schools, the Education Park of Anyi County is one of the first education parks of county level approved by the State. It becomes the major design conception to establish the cultural and ecological axis of the Park as the planned structural framework. The exterior wall adopts brick red as key tone combining light-weight steel structure and transparent material, the contemporary and stylish shape shows sculptural beauty and cultural deposits. The architecture is arranged around the cultural and ecological axis of the campus in south-north direction. At the center of the ecological axis there lies public buildings such as library, Youth Activity Center and Gymnasium etc., which serve as the resource sharing area for sever schools.

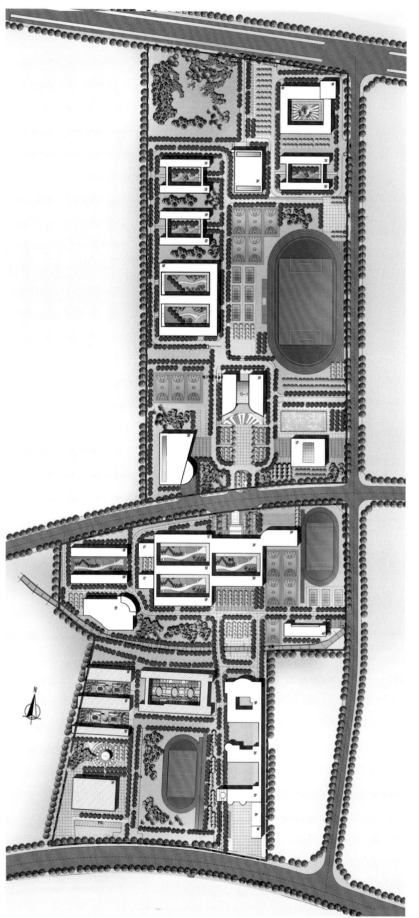

▲规划设计总平面图 Master Planning + Urban Design

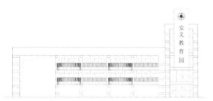

▲ 立面图 *Elevation*

▲ 立面图 *Elevation*

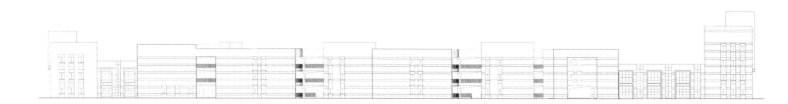

▲ 剖面图 *Saction*

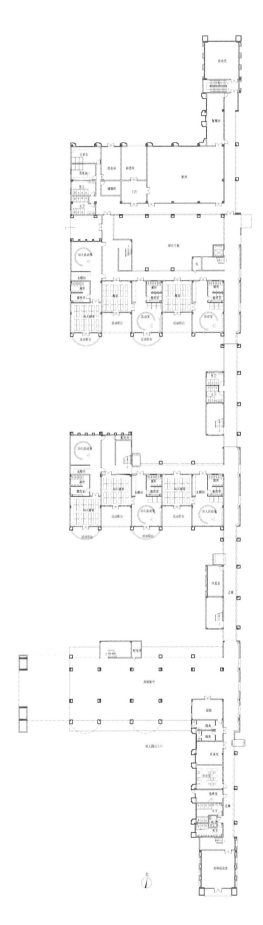

▲ 一层平面图 Level One Plan

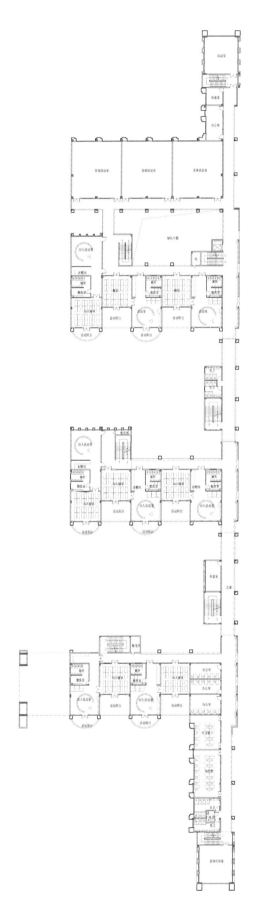

▲ 二层平面图 Level Two Plan

▲ 教学楼立面图 Elevation of Educational Building　　　　　　▲ 教学楼立面图 Elevation of Educational Building

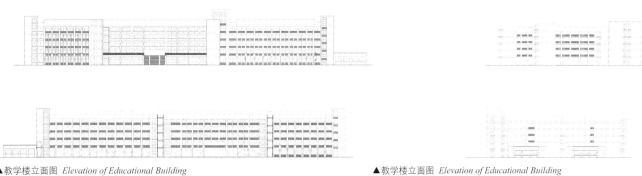

▲ 教学楼一层平面图 Level One Plan of Educational Building　　　▲ 教学楼二层平面图 Level Two Plan of Educational Building

桂林航天工业学院来宾校区

Laibin Campus of Guilin University of Aerospace Technology

项目地点： Venue:	广西省来宾市 Laibin City, Guangxi Province
设计日期： Design Date:	2013 年 2013
用地面积： Project Area:	480,000 平方米 480,000m²
项目建筑面积： Construction Area:	670,000 平方米 670,000m²
合作设计： Partner:	冯险峰、黄军鹏、李滔、李丹平 Feng Xianfeng, Huang Junpeng, Li Tao, Li Danping

桂林航天工业学院来宾校区占地 48 万平方米，是以航空专业为龙头的大专院校，用地位于来宾市新区中心，有城市观光河道穿越校园。

本校区规划结构呈"一心二轴"，"一心"指铜鼓造型圆形的图书馆，"二轴"指贯穿校园南北的校园文化生态轴和城市水轴，其文化生态轴将校园教学区、生活区、体育运动区连成一体，在收放有序的生态轴上，以"迎天、问天、观天、飞天"为主题，创造航天院校独特的校园文化氛围，并为师生提供一系列交流、休闲的学习场所，以突出航天院校文化及生态这一基本主题。

Laibin Campus of Guilin University of Aerospace Technology is specialized in professional aerospace in downtown of Laibin City of 48 hectares. There is sightseeing channel of the city running through the campus.

The structure is planned based on "one core, two axis", which "one core" refers to the circular library in the shape of brass drum, while "two axis" indicates the cultural ecological axis and city waterscape axis penetrating south and north of the campus. The Classroom Area, Living Quarters and Sports Area are linked together on the flexible and ordered cultural ecological axis. Creating the theme of "welcoming, asking, observing and flying the sky", unique cultural ambience is engendered, and a series of communication, recreation and study space is provided to give prominence to the basic theme of culture and ecology of Guilin University of Aerospace Technology.

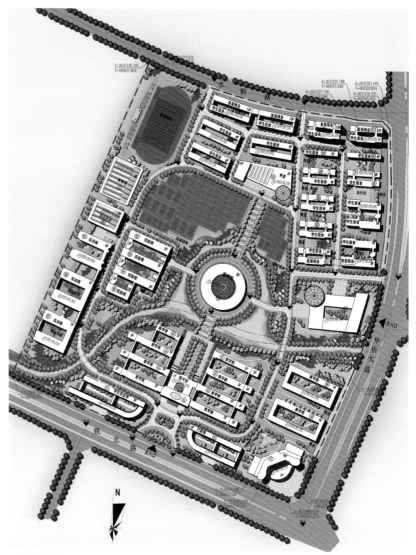

▲规划设计总平面图 *Master Planning + Urban Design*

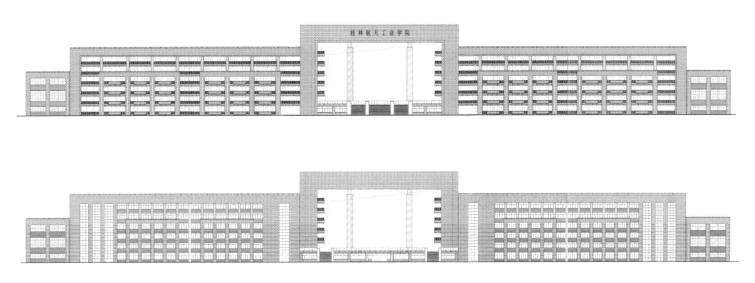

▲ 立面图 *Elevation*

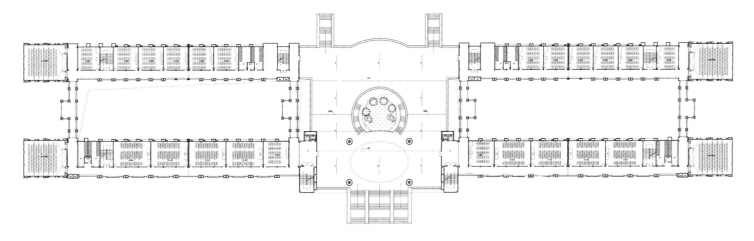

▲ 一层平面图 *Level One Plan*

广州市增城太阳城巧克力社区
Chocolate Community of Sun City at Zengcheng, Guangzhou

清远市凯旋门商住区
Commercial & Residential District of Triumphal arch, Qingyuan

南宁市盛天绿都住宅区
Residential District of Shengtian Lüdu, Nanning

赣州市云星中央星城
Yunxing Central Star Building, Ganzhou

太原市东山雅居住宅区
Commercial & Residential District of Dongshan Yaju, Taiyuan

广州市花城湾畔
HuaCheng Bay, Guangzhou

太原市龙湾国际商住区
Commercial & Residential District of Longwan International, Taiyua

南宁市钱隆天下商住区规划建筑设计
Planning & Architecture Design of Commercial & Residential District of Qianlong Tianxia, Nanning

江门市开平云景长青路住宅项目
Residential District of Kaiping Yunjing Changqing Lu, Jiangmen

住宅 RESIDENTIAL

广州市萝岗峻森、锦泽商住区规划
Planning of Commercial & Residential District of Luogang, Guangzhou

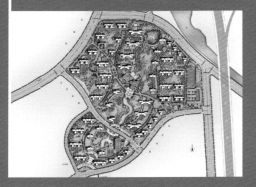

南宁市钱隆御景商住区规划与建筑设计
Planning & Architecture Design of Commercial & Residential District of Qianlong Yujing, Nanning

广州亚运会亚运村住宅东区规划
Planning of East Residential District for Guangzhou Asian Games, Guangzhou

重庆南坪万达广场
Nanping Wanda Plaza, Chongqing

重庆金科·廊桥水乡
Jinke · The Bridges of Riverside Town, Chongqing

广州市万德福
Wandefu, Guangzhou

南宁市城市春天
City Spring, Nanning

清远市云景中汇园
Yunjing Zhonghuiyuan, Qingyuan

广州市花都区俊怡利丰住宅项目
Residential District of Junyilifeng, Huadu, Guangzhou

广州市花都雅图奥园
Huadu Yatu Aoyuan, Guangzhou

广州市广东华商学院教师住宅
Residential District of Guangdong Huashang College, Guangzhou

惠州市文头岭住宅区
Residential District of Touwenling, Huizhou

晋中市榆次小南庄住宅区规划
Planning of Xiaonanzhuang Residential District, Yuci, Jinzhong

防城港市东兴金滩英伦海
Planning of Commercial & Residential District of Dongxing JintanYinglunhai, Fangchenggang

广州珠江公园旁高层大宅
High-rise Residence by Zhujiang Park (Interior Design), Guangzhou

广恒·世纪城
Guangheng · Century City

山西运城市珍爱佳苑小区
Residential District of Zhenaijiayuan, Yuncheng, Shanxi

广西裕丰荔园
Yufeng Liyuan, Guangxi

后记
POSTSCRIPT

本书内容甄选自建筑师黄劲先生的数十项代表设计作品，涉及酒店建筑、商业建筑、教育建筑、公共建筑、住宅建筑以及城市设计，其中不乏荣获国家、省、市及行业协会奖项的作品。通过对不同类型的案例品评，读者可以全面而直观地领略到黄劲先生对新岭南建筑创作的思想，感悟其融合中西建筑科技，延续百年岭南现代建筑设计神韵及表现形式的苦心，更可以从中体会到中国职业建筑师的探索与创新之成长历程。

能有机会承担专辑的编辑、摄影工作，有赖作者的信赖与支持，使我最大程度地自由进行编辑与版面设计，整理相关的项目资料，近距离观察各项目作品，更唤醒和更新了我在酒店管理、商业地产、设计及技术等相关方面的经验和知识。

借着本书，也算是部分实现了我们 AEC 建筑＋设计沙龙联展与论坛的多年夙愿，为岭南建筑设计的优秀代表们编撰其个人作品集，发掘第一手资料去展现新岭南建筑设计的现状与发展，不遗余力地推动其发扬光大，以给予大家更多的启发与协助。

借此，向各岭南建筑设计的先行者致敬！

叶飚
于 2014 年秋

Tens of signature designs are well-chosen from Mr. Huang Jin's works including hotel, commercial, educational, public, residential and urban architectures, among which there are a number of award-winning projects at the level of industry, city, province and state. Through the comment on different cases, the Readers could comprehensively appreciate Mr. Huang Jin's Neo-Lingnan architecture creation, and his painstaking effort of integrating Chinese and Western architecture technology to preserve the design essence and expression of Lingnan Style with a history of over a hundred years, as well understand the growth of exploration and creation of the professional architects in China.

Thanks for the trust and support from Mr. Huang Jin, that I am honored to take the chance of being the editor and photographer of the Book. I am able to freely edit, graphic-design and organize the relevant project materials. By closely observing the projects, my pertinent experience and knowledge is aroused and updated in regard to hotel management, commercial property, design and technology etc. .

Owing to this Book, the long-cherished wish is also partly fulfilled for our AEC Architecture + Design Salon Joint Exhibition and Forum: to compile personal work collections for excellent designers of Lingnan architecture, to exploit and present the current status and development of Neo-Lingnan architecture design and go all out to carry it forward, to give more inspiration and assistance to those who need it.

I would like to take this opportunity to pay my respects to the pioneers of all Lingnan architecture designers!

Billy Yip
2014 Autumn

图书在版编目（CIP）数据

筑梦天下——黄劲 / 黄劲、黄思宇编著；叶飚摄影 . -- 沈阳：辽宁科学技术出版社，2015.2
　ISBN 978-7-5381-8993-3

　Ⅰ. ①筑… Ⅱ. ①黄… ②黄… ③叶… Ⅲ. ①建筑设计—作品集—中国—现代 Ⅳ. ① TU206

中国版本图书馆 CIP 数据核字 (2015) 第 008366 号

出 版 发 行：	辽宁科学技术出版社
	（沈阳市和平区十一纬路 29 号　邮编：110003）
经 销 者：	各地新华书店
印 刷 者：	广州市恒远彩印有限公司
版 面 设 计：	广州市弘志广告设计有限公司
策 划 者：	维捷 - 机域设计传播顾问公司
幅面尺寸：	235mm×280mm
总 印 张：	16
插　　页：	4
字　　数：	360 千字
出版时间：	2015 年 2 月第 1 版
印刷时间：	2015 年 2 月第 1 次印刷
特邀编辑：	叶　飚
翻　　译：	黄健茵　何　炜
责任编辑：	郭　健
装帧设计：	姚懿恩
制　　作：	叶仲轩 陈富瑶 张景致 姚懿恩 邝家辉
责任校对：	魏春爱
书　　号：	ISBN 978-7-5381-8993-3
定　　价：	228.00 元

联系编辑：024-23284536，13898842023
邮购热线：024-23284502
E-mail:1013614022@qq.com

点线间融合中西建筑科技，延续百年岭南现代建筑设计神韵

The Dotted Line Fusion of Chinese and Western Architectural Technology, Continuation of Centuries Lingnan Genre of Architectural Design Charm.